Division of Psychiatry
University of Chicago
Chicago, Illinois

LIST OF VOLUMES PUBLISHED IN THIS SERIES

(Some of these are temporarily out of print.)

PHILOSOPHY

1. **First and Last Things.**	H. G. Wells.
3. **The Riddle of the Universe.**	Ernst Haeckel.
5. **On Liberty.**	J. S. Mill.
57. **Clearer Thinking : Logic for Everyman.**	A. E. Mander.
62. **First Principles.**	Herbert Spencer.
78. **The Man versus the State.**	Herbert Spencer.
84. **Let the People Think.**	Bertrand Russell.
101. **Flight from Conflict.**	Laurence Collier.
112. **Human Nature, War, and Society.**	Dr. John Cohen.
113. **The Rational Good.**	L. T. Hobhouse.
116. **The Illusion of National Character.**	Hamilton Fyfe.
119. **Ideals and Illusions.**	L. Susan Stebbing.
125. **Man His Own Master.**	Archibald Robertson.

PSYCHOLOGY

46. **The Mind in the Making.**	James Harvey Robinson.
48. **Psychology for Everyman (and Woman).**	A. E. Mander.
85. **The Myth of the Mind.**	Frank Kenyon.
115. **The Distressed Mind.**	J. A. C. Brown, M.B., Ch.B.
117. **Population, Psychology, and Peace.**	Prof. J. C. Flugel.
122. **The Evolution of Society.**	J. A. C. Brown.

ANTHROPOLOGY

14 & 15. **Anthropology** (2 vols.).	Sir E. B. Tylor.
26. **Head-hunters : Black, White, and Brown** (illus.).	Dr. A. C. Haddon.
29. **In the Beginning : The Origin of Civilization** (illus.).	Prof. Sir G. Elliot Smith.
40. **Oath, Curse, and Blessing.**	Ernest Crawley.
45. **Men of the Dawn** (illus.).	Dorothy Davison.
87. **Man Makes Himself.**	V. Gordon Childe.
102. **Progress and Archæology.**	V. Gordon Childe.
127. **The Earliest Englishman** (illus.).	Sir Arthur Smith Woodward.

SCIENCE

12. **The Descent of Man.**	Charles Darwin.
36. **Savage Survivals** (illus.).	J. Howard Moore.
41. **Fireside Science.**	Sir E. Ray Lankester.
47. **The Expression of the Emotions in Man and Animals** (illus.).	Charles Darwin.
59. **Your Body : How it is Built and How it Works** (illus.).	Dr. D. Stark Murray.
61. **Man and His Universe.**	John Langdon-Davies.
65. **Dictionary of Scientific Terms.**	C. M. Beadnell, C.B., F.Z.S.
67. **The Universe of Science.**	Prof. H. Levy.
89. **The Origin of the Kiss, and other Scientific Diversions.**	C. M. Beadnell, C.B., F.Z.S.
94. **Life's Unfolding.**	Sir Charles Sherrington, O.M.
95. **An Easy Outline of Astronomy.**	Dr. Martin Davidson.
97. **Man Studies Life.**	G. N. Ridley.
103. **The Chemistry of Life.**	J. S. D. Bacon, M.A.
104. **Medicine and Mankind.**	Dr. A. Sorsby.
108. **Geology in the Life of Man.**	Duncan Leitch.
114. **Man : The Verdict of Science.**	G. N. Ridley.
120. **An Outline of the Development of Science** (illus.)	Mansel Davies.
123. **Background to Modern Thought.**	C. D. Hardie.
128. **Astronomy for Beginners.** (illus.).	Dr. Martin Davidson.
129. **The Search for Health.** (illus.).	Dr. D. Stark Murray.

RELIGION

No.	Title	Author
4.	**Humanity's Gain from Unbelief, and other selections from the Works of Charles Bradlaugh.**	
9.	**Twelve Years in a Monastery.**	JOSEPH MCCABE.
11.	**Gibbon on Christianity.**	
17.	**Lectures and Essays.**	T. H. HUXLEY.
18.	**The Evolution of the Idea of God.**	GRANT ALLEN.
19.	**An Agnostic's Apology.**	Sir LESLIE STEPHEN, K.C.B.
24.	**A Short History of Christianity.**	J. M. ROBERTSON.
30.	**Adonis : A Study in the History of Oriental Religion.**	Sir J. G. FRAZER.
34.	**The Existence of God.**	JOSEPH MCCABE.
44.	**Fact and Faith.**	Prof. J. B. S. HALDANE.
49.	**The Religion of the Open Mind.**	A. GOWANS WHYTE.
51.	**The Social Record of Christianity.**	JOSEPH MCCABE.
52.	**Five Stages of Greek Religion.**	Prof. GILBERT MURRAY, O.M.
53.	**The Life of Jesus.**	ERNEST RENAN.
54.	**Selected Works of Voltaire.**	Trans. by JOSEPH MCCABE.
69.	**The Age of Reason.**	THOMAS PAINE.
83.	**Religion Without Revelation.**	JULIAN S. HUXLEY.
90 & 91.	**The Bible and Its Background** (2 vols.).	ARCHIBALD ROBERTSON.
93.	**The Gospel of Rationalism.**	C. T. GORHAM.
96.	**The God of the Bible.**	EVANS BELL.
98.	**In Search of the Real Bible.**	A. D. HOWELL SMITH.
99.	**The Outlines of Mythology.**	LEWIS SPENCE.
100.	**Magic and Religion.**	Sir J. G. FRAZER.
105.	**The Church and Social Progress.**	MARJORIE BOWEN.
106.	**The Great Mystics.**	GEORGE GODWIN.
107.	**The Religion of Ancient Mexico.**	LEWIS SPENCE.
109.	**A Century for Freedom.**	Dr. KENNETH URWIN.
110.	**Jesus : Myth or History ?**	ARCHIBALD ROBERTSON.
111.	**The Ethics of Belief, and Other Essays.**	W. K. CLIFFORD.
124.	**The Holy Heretics.**	EDMOND HOLMES.
126.	**Men Without Gods.**	HECTOR HAWTON.
132.	**The Origins of Religion.**	Lord RAGLAN.

HISTORY

No.	Title	Author
6.	**A Short History of the World** (revised to 1946).	H. G. WELLS.
13.	**History of Civilization in England** (Vol. 1).	H. T. BUCKLE.
25.	**The Martyrdom of Man.**	WINWOOD READE.
39.	**Penalties Upon Opinion.**	H. BRADLAUGH BONNER.
72.	**A Short History of Women.**	JOHN LANGDON-DAVIES.
121.	**Head and Hand in Ancient Greece.**	PROF. B. FARRINGTON.

FICTION

No.	Title	Author
118.	**Friar's Lantern.**	Dr. G. G. COULTON.
130.	**The Mystery of Anna Berger.**	GEORGE GODWIN.

MISCELLANEOUS

No.	Title	Author
2.	**Education : Intellectual, Moral, and Physical.**	HERBERT SPENCER.
7.	**Autobiography of Charles Darwin.**	
16.	**Iphigenia.**	Two Plays by EURIPIDES.
28.	**The City of Dreadful Night, and Other Poems.**	JAMES THOMPSON ("B.V.")
32.	**On Compromise.**	JOHN VISCOUNT MORLEY, O.M., P.C.
43.	**The World's Earliest Laws.**	CHILPERIC EDWARDS.
60.	**What is Man ?**	MARK TWAIN.
63.	**Rights of Man.**	THOMAS PAINE.
64.	**This Human Nature.**	CHARLES DUFF.
71.	**A Candidate for Truth.** Passages from RALPH WALDO EMERSON chosen and arranged by	GERALD BULLETT.
74.	**Morals, Manners, and Men.**	HAVELOCK ELLIS.
75.	**Pages from a Lawyer's Notebooks.**	E. S. P. HAYNES.
79.	**The World as I See It.**	ALBERT EINSTEIN.
86.	**The Liberty of Man, and Other Essays.**	R. G. INGERSOLL.
131.	**Wrestling Jacob.**	MARJORIE BOWEN.

Frontispiece.

CHARLES DARWIN

The Thinker's Library, No. 47 (Science Section).

THE EXPRESSION OF THE EMOTIONS IN MAN AND ANIMALS

BY

CHARLES DARWIN

REVISED AND ABRIDGED BY

SURGEON-REAR-ADMIRAL C. M. BEADNELL

C.B., K.H.P. M.R.C.S. (ENG.)

LATE FELLOW OF THE CHEMICAL SOCIETY AND THE ROYAL ANTHROPOLOGICAL INSTITUTE

LONDON:

WATTS & CO.,

5 & 6 JOHNSON'S COURT, FLEET STREET, E.C.4

First published in the Thinker's Library,
1934
Second Impression, 1943
Third Impression, 1948

THE PAPER (AND BINDING) OF THIS BOOK CONFORMS TO THE AUTHORISED ECONOMY STANDARDS

Printed and Published in Great Britain by C. A. Watts and Co. Limited,
5 & 6 Johnson's Court, Fleet Street, London, E.C.4

FOREWORD

Abraham. Do you bite your thumb at us, sir?
Sampson. No, sir, I do not bite my thumb at you sir; but I bite my thumb, sir.

Romeo and Juliet.

DARWIN wrote the *Expression of the Emotions* before physiologists had discovered the astonishing facts that the innermost moods of the mind—love, hate, fear, rage, etc.—together with their outermost manifestations, the muscular contractions of the limbs and face, are in great measure dependent upon the functioning, in appropriate spheres of action, of "chemical messengers," or *hormones*, which are despatched from specific "chemical factories" known as the endocrine glands. The discovery would have rejoiced Darwin's heart, since it lends strong support to the deterministic physico-chemical, as opposed to the dualistic teleological, interpretation of biological phenomena. Despite these radical changes the main thesis in this classic in physiognomy is as irrefutable to-day as when it was written, thanks to the enormous number of observations upon which Darwin's inductions were based, and the scrupulous care with which he eliminated the personal equation. This greatest of naturalists and kindliest of men was not only a man of science but, in addition, that *rara avis*, a scientist self-disciplined to consecutive and consistent thought. He fearlessly grasped with both hands knowledge founded on experience, and did not content himself, as do certain scientists, with holding on to the empirical by one hand while clutching at the mystical with the other.

In the re-editing of the original work in abbreviated form, the wealth of evidence Darwin amassed to corroborate his arguments has in one respect made the task easy, though it has involved the difficulty of deciding what must be sacrificed in a text every sentence of which teems with interest.

Many of the curious, yet familiar, "mannerisms," the nature and significance of which are herein described, are by no means of academic interest only; they are of practical, maybe even of vital, import. Knowledge is power, and all who are intimately concerned with the reading of the mind of man and beast—artists, actors, physicians, teachers, judges, farmers, and big-game hunters—should be in a position at once to sense the mental state lying behind this or that gesticulation, gesture, or pose. Let me endorse this contention with three illustrations—a utilitarian, a comic, and a tragic.

A man desirous of buying a bull asked to see his teeth, but nothing would persuade the animal to show them. At last someone versed in taurine behaviour suggested that a cow should be brought. This was no sooner done than the bull stretched out his neck, everted his lips, and, in a broad grin, exposed his teeth.

A little boy once bet his chum he would "beat the band." Placing himself in the forefront of the players he proceeded, to the accompaniment of apposite facial contortions, to suck a lemon. The consequent association of ideas called up in the minds of the performers paralysed their salivary glands and dried up their mouths, so that after a few desperate but discordant blasts and squeals the music ceased for sheer want of lubrication.

In the war of 1914 a British naval officer was shot by the Turks off the coast of Kuşadasi. This was apparently due to the fact (which Mr. Sedat Zeki of the Turkish Embassy in London has kindly corroborated) that the Turkish "Come here! Advance!" gesture so closely resembles our "Go away! Retire!" gesture that it was mistaken for it, with fatal result, by the English officers in the boat.

CHARLES M. BEADNELL.

FOREWORD
TO THIRD IMPRESSION

By MAURICE BURTON, D.Sc.

THE Theory of Evolution, The Descent of Man, and The Struggle for Existence have become almost household words, and the association of Darwin's name with them, however vague, is irrevocably fixed in the minds of most people. Yet, despite this familiarity with the name of Darwin, our acceptance of his scientific eminence is apt to be based upon his alleged enunciation of the theory that man is descended from a monkey. It is, indeed, a lamentable thing that a man who was outstanding in his painstaking search for truth should have his memory perpetuated in the popular mind almost entirely by reason of a distortion of it! Nor can we blame the layman overmuch, for even among professional biologists his works are seldom given the attention they deserve.

In reading his *Expressions of the Emotions*, as indeed in reading any of his works, one is struck by the extreme simplicity of his observations. There is little in this book which the ordinary man could not observe for himself; there is little in the deductions which the ordinary man could not arrive at for himself—once it has been pointed out! Therein we have the secret of Darwin's greatness; and it is always worth while to read or re-read his works to rediscover the man, as well as to learn something of his ideas. Nowadays, in a world in which there is a return to the study of biology in the field, it is important to remember that Darwin was our greatest field-naturalist, who, by the acuteness of his observations of simple things, explained in homely terms, profoundly transformed the philosophy of the civilised world.

The modern trend in biology is towards an increasingly greater stress on the behaviour of animals, and particularly towards the study of animals in their natural surroundings. In other words, it is becoming noticeably popular to forsake the stuffed skin, the mounted skeleton, and the animal preserved in alcohol. Instead there is the desire to go out and observe the animal as a living being. Although we have progressed considerably in such studies since Darwin wrote, he was a pioneer and the modern works on animal behaviour can be the more readily understood by re-tracing the early trail he blazed.

The Expression of the Emotions is a classic. One might almost say, in addition, a museum-piece; and it is interesting to note how its author, as, for example, in the paragraph starting at the middle of page 43, applies so sparingly and almost hesitatingly his own Theory of Natural Selection, whereas to-day it is a cardinal principle assumed and accepted as the only possible explanation in such cases.

Finally, it may be asked: wherein lies the importance of Darwin's work, and why need we concern ourselves with animal behaviour and emotions? True, the subject is interesting, but is it any more than this? The simple answer is that unless and until it is more generally realized that man has a oneness with the animal kingdom in the expression of his emotions, and, whether we will or no, that the greater part of human behaviour is basically emotional, our approach to many urgent social problems will remain distorted.

CONTENTS

CHAP.		PAGE
	Introduction	1
I.	General Principles of Expression	4
II.	General Principles of Expression (*continued*)	14
III.	General Principles of Expression (*concluded*)	24
IV.	Means of Expression in Animals	33
V.	Special Expressions of Animals	54
VI.	Special Expressions of Man: Suffering and Weeping	72
VII.	Low Spirits, Anxiety, Grief, Dejection, Despair	87
VIII.	Joy, High Spirits, Love, Tender Feelings, Devotion	98
IX.	Reflection — Meditation—Ill-temper—Sulkiness—Determination	110
X.	Hatred and Anger	118
XI.	Disdain — Contempt — Disgust — Guilt—Pride, etc.—Helplessness — Patience — Affirmation and Negation	127
XII.	Surprise — Astonishment — Fear — Horror	140
XIII.	Self-attention—Shame—Shyness—Modesty: Blushing	153
XIV.	Concluding Remarks and Summary	166
	Index	173

LIST OF ILLUSTRATIONS

FIG.		PAGE
	Charles Darwin	*Frontispiece*
1.	Dog approaching with hostile intentions	14
2.	Dog in a humble and affectionate frame of mind	15
3.	Domestic Cat, savage and prepared to fight	17
4.	Domestic Cat in an affectionate frame of mind	18
5.	The Lynx (Caracal) with hate and fury boiling in his breast	19
6.	Domestic Cat threatening a dog	20
7.	Wild (feral) Cat from Scotland	30
8.	Lioness and Cub	31
9.	Hen angry and driving away a dog from her chickens	40
10.	Enraged Swan driving away an intruder	41
11.	Head of snarling Dog	49
12.	A Baboon (*Chacma* or *Cynocephalus*) showing off his dental armature	52
13.	Human-like expression in young Chimpanzee	64
14.	Chimpanzee disappointed and sulky	68
15.	Little inmates of a Children's Home	82
16.	Small Boy " down in the mouth "	94
17.	Same Boy later	95
18.	Sneering and defiance	124
19.	The expression is that of disdain	128
20.	Disgust	130
21.	Terror	145
22.	Horror and Pain	150

THE EXPRESSION

OF THE

EMOTIONS IN MAN AND ANIMALS

INTRODUCTION

SIR CHARLES BELL, so illustrious for his discoveries in physiology, may be said not only to have laid the foundations of the anatomy and philosophy of "Expression" as a branch of science, but to have built up a noble structure. His service consists chiefly in having shown the intimate relation which exists between the movements of expression and those of respiration. One most important point, small as it may at first appear, is that the muscles round the eyes are involuntarily contracted during violent expiratory efforts, in order to protect these delicate organs from the pressure of the blood. This fact throws a flood of light on several of the most important expressions of the human countenance.

Mr. Herbert Spencer makes the following remarks: "Fear, when strong, expresses itself in cries, in efforts to hide or escape, in palpitations and tremblings; and these are just the manifestations that would accompany an actual experience of the evil feared. The destructive passions are shown in a general tension of the muscular system, in gnashing of the teeth, and protrusion of the claws, in dilated eyes and nostrils, in growls; and these are weaker forms of the actions that accompany the killing of prey." Here we have, as I believe, the true theory of a large number of expressions. Mr. Spencer insists on "the general law that feeling passing a certain pitch, habitually vents itself in bodily action"; and that "an overflow of nerve-force undirected by any motive will manifestly take first the most habitual routes; and if these do not suffice, will next overflow into the less habitual ones." This law I believe to be of the highest importance in throwing light on our subject.

All the authors who have written on Expression, with the exception of Mr. Spencer—the great expounder of the principle of Evolution—appear to have been firmly convinced that species, man of course included, came into existence in their present condition. Sir C. Bell maintains that many of our facial muscles are "a special provision" for this sole object of expression. But the simple fact that the anthropoid apes possess the same facial muscles as we do renders it very improbable that these muscles in our case serve exclusively for expression; for no one, I presume, would be inclined to admit that monkeys have been endowed with special muscles solely for exhibiting grimaces. Distinct uses, independently of expression, can be assigned for almost all the facial muscles.

Man himself cannot express love and humility by external signs so plainly as does a dog when, with drooping ears, hanging lips, flexuous body, and wagging tail, he meets his beloved master. Nor can these movements in the dog be explained by acts of volition or necessary instincts, any more than the beaming eyes and smiling cheeks of a man when he meets an old friend. Sir C. Bell would no doubt have said that the dog had been created with special instincts, adapting him for association with man.

As long as man and all other animals are viewed as independent creations, an effectual stop is put to our natural desire to investigate the causes of Expression. By this doctrine anything and everything can be equally well explained; and it has proved as pernicious with respect to Expression as to every other branch of natural history. With mankind some expressions, such as the bristling of the hair under the influence of extreme terror, or the uncovering of the teeth under that of furious rage, can hardly be understood except on the belief that man once existed in a much lower and animal-like condition. The movement of the same facial muscles during laughter by man and by various monkeys is rendered more intelligible if we believe in their descent from a common progenitor. He who admits on general grounds that the structure and habits of all animals have been gradually evolved will look at the whole subject of Expression in a new and interesting light.

In the year 1838 I was already inclined to believe in the principle of evolution, or of the derivation of species from other and lower forms. Consequently Sir C. Bell's view, that man had been *created* with certain muscles specially adapted for the expression of his feelings, struck me as unsatisfactory. It seemed probable that the habit of expressing our feelings by certain movements had been in some manner gradually acquired, and that each expression demanded a *rational* explanation.

CHAPTER I

GENERAL PRINCIPLES OF EXPRESSION

I WILL begin by giving the three Principles which appear to me to account for most of the expressions and gestures involuntarily used by man and the lower animals under the influences of various emotions and sensations.[1] Movements or changes in any part of the body—as the wagging of a dog's tail, the drawing back of a horse's ears, the shrugging of a man's shoulders, or the dilatation of the capillary vessels of the skin—may all equally well serve for expression.

I. *The Principle of Serviceable Associated Habits.*—Certain complex actions are of direct or indirect service under certain states of the mind, in order to relieve or gratify certain sensations, desires, etc.; and whenever the same state of mind is induced, however feebly, there is a tendency through the force of habit and association for the same movements to be performed, though they may not then be of the least use. Some actions ordinarily associated through habit with certain states of the mind may be partially repressed through the will, and in such cases the muscles which are least under the separate control of the will are the most liable still to act, causing movements which we recognize as expressive. In certain other cases the checking of one habitual movement requires other slight movements; and these are likewise expressive.

II. *The Principle of Antithesis.*—Certain states of

[1] Mr. Herbert Spencer has drawn a clear distinction between emotions and sensations, the latter being "generated in our corporeal framework." He classes as feelings both emotions and sensations.

the mind lead to certain habitual actions which are of service. Now, when a directly opposite state of mind is induced, there is a strong and involuntary tendency to the performance of movements of a directly opposite nature, though these are of no use; and such movements are in some cases highly expressive.

III. *The Principle of the Direct Action of the Nervous System.*—When the sensorium is strongly excited nerve-force is generated in excess, and is transmitted in certain definite directions, depending on the connection of the nerve-cells and partly on habit; or the supply of nerve-force may be interrupted. Effects are thus produced which we recognize as expressive.

It is notorious how powerful is the force of habit. The most complex and difficult movements can in time be performed without the least effort or consciousness. Habit is efficient in facilitating complex movements; physiologists admit that the conducting power of the nervous fibres increases with the frequency of their excitement. This applies to nerves of motion and sensation, as well as to those connected with the act of thinking. That some physical change is produced in the nerve-cells and nerves which are habitually used can hardly be doubted.

When there exists an inherited or instinctive tendency to the performance of an action, or an inherited taste for certain kinds of food, some degree of habit in the individual is often or generally requisite. We find this to a certain extent in the pointing of dogs; although some young dogs point excellently the first time they are taken out, yet they often associate the proper inherited attitude with a wrong odour, and even with eyesight. Caterpillars which have been fed on the leaves of one kind of tree have been known to perish from hunger rather than eat the leaves of another tree, although this afforded them their proper food under a state of nature.

Mr. Bain remarks that "actions, sensations, and states of feeling, occurring together or in close succession, tend to grow together, or cohere, in such a way that when any one of them is afterwards presented to the mind the others are apt to be brought up in idea."[1] It is known to every one how difficult or even impossible it is, without repeated trials, to move the limbs in certain opposed directions which have never been practised. Analogous cases occur with sensations, as in the common experiment of rolling a marble beneath the tips of two crossed fingers, when it feels exactly like two marbles. Every one protects himself when falling to the ground by extending his arms, and few can resist acting thus when voluntarily falling on a soft bed.

When our minds are much affected, so are the movements of our bodies; but here another principle besides habit—namely, the undirected overflow of nerve-force—partially comes into play.

> "Some strange commotion
> Is in his brain: he bites his lip, and starts;
> Stops on a sudden, looks upon the ground,
> Then lays his finger on his temple; straight
> Springs out into fast gait; then stops again,
> Strikes his breast hard, and anon he casts
> His eye against the moon: in most strange postures
> We have seen him set himself."
>
> —*Hen. VIII*. Act iii. sc. 2.

A man often scratches his head when perplexed; and I believe he acts thus from habit, as if he experienced a slightly uncomfortable bodily sensation—namely, the itching of his head. Another man rubs his eyes when perplexed, or gives a little cough when embarrassed, acting in either case as if he felt a slightly uncomfortable sensation in his eyes or windpipe.

From the continued use of the eyes, these organs

[1] Professor Huxley remarks: "It may be laid down as a rule that, if any two mental states be called up together, or in succession, with due frequency and vividness, the subsequent production of the one of them will suffice to call up the other, and that whether we desire it or not."

are especially liable to be acted on through association under various states of the mind, although there is manifestly nothing to be seen. A man who vehemently rejects a proposition will almost certainly shut his eyes or turn away his face; but if he accepts the proposition he will nod his head in affirmation and open his eyes widely. The man acts in this latter case as if he clearly saw the thing, and in the former case as if he did not or would not see it. Persons in describing a horrid sight often shut their eyes momentarily and firmly, or shake their heads, as if not to see or to drive away something disagreeable; and I have caught myself, when thinking in the dark of a horrid spectacle, closing my eyes firmly. In looking suddenly at any object, or in looking all around, every one raises his eyebrows, so that the eyes may be quickly and widely opened; and a person in trying to remember something often raises his eyebrows, as if to see it.

There are other actions which seem to be due to imitation or some sort of sympathy. Thus persons cutting anything with a pair of scissors may be seen to move their jaws simultaneously with the blades of the scissors. Children learning to write often twist about their tongues as their fingers move, in a ridiculous fashion. When a public singer suddenly becomes a little hoarse many of those present may be heard to clear their throats; but here habit probably comes into play, as we clear our own throats under similar circumstances.

Reflex Actions.—Reflex actions are due to the excitement of a peripheral nerve, which transmits its influence to certain nerve-cells, and these in their turn excite certain muscles or glands to action; and all this may take place without any sensation or consciousness on our part, though often thus accompanied. Some of them graduate into and can hardly be distinguished from actions which have arisen through habit. Coughing and sneezing are familiar instances of reflex actions. With infants the first act of respiration is often a sneeze, although this requires the co-ordinated movement of numerous muscles. Respiration is partly voluntary, but mainly reflex, and is performed in the most natural and best manner without the interference of the will. A vast number of complex movements are reflex. As good an instance as can be given is the one of a decapitated frog, which cannot of course feel, and cannot consciously perform any movement. Yet if a drop of acid be placed on the lower surface of the thigh of a frog in this state it will rub off the drop with the upper surface of the foot of the same leg. If this foot be cut off it cannot thus act. After some fruitless efforts, therefore, it gives up trying in that way, and at last makes use of the foot of the other leg and succeeds in rubbing off the acid. We have here co-ordinated contractions in due sequence for a special purpose—contractions that " have all the appearance of being guided by intelligence and instigated by will in an animal, the recognized organ of whose intelligence and will has been removed."[1]

We see the difference between reflex and voluntary movements in very young children not being able to blow their noses and not being able to clear their throats of phlegm. They have to learn to perform these acts, yet they are performed, when a little older, almost as easily as reflex actions. Sneezing and coughing, however, can be controlled by the will only partially or not at all; whilst the clearing the throat and blowing the nose are completely under our command.

The conscious wish to perform a reflex action sometimes stops or interrupts its performance, though the proper sensory nerves may be stimulated. For instance, I laid a small wager with a dozen young men that they would not sneeze if they took snuff, although they all declared that they invariably did so; accordingly they all took a pinch, but from wishing much to succeed not one sneezed, though their eyes watered,

[1] Dr. Maudsley, *Body and Mind.*

and all, without exception, had to pay me the wager. Attention paid to the act of swallowing interferes with the proper movements; from which it probably follows that some persons find it so difficult to swallow a pill.

Another familiar instance of a reflex action is the involuntary closing of the eyelids when the surface of the eyes is touched. A similar winking movement is caused when a blow is directed towards the face. The whole body and head are generally at the same time drawn suddenly backwards. These latter movements, however, can be prevented if the danger does not appear imminent; but our reason telling us that there is no danger does not suffice. I put my face close to the thick glass-plate in front of a puff-adder in the Zoological Gardens, with the firm determination of not starting back if the snake struck at me; but as soon as the blow was struck my resolution went for nothing, and I jumped a yard or two backwards with astonishing rapidity. My will and reason were powerless against the imagination of a danger which had never been experienced.

A start from a sudden noise, when the stimulus is conveyed through the auditory nerves, is always accompanied in grown-up persons by the winking of the eyelids. Though my infants under a fortnight old started at sudden sounds, they certainly did not always wink their eyes, and I believe never did so. The start of an older infant apparently represents a vague catching hold of something to prevent falling. I shook a pasteboard box close before the eyes of one of my infants, when 114 days old, and it did not wink; but when I put a few comfits into the box, holding it in the same position as before, and rattled them, it blinked its eyes violently every time, and started a little.

It appears probable that starting was originally acquired by the habit of jumping away as quickly as possible from danger, whenever any of our senses gave us warning. Starting, as we have seen, is accompanied by the blinking of the eyelids so as to protect the eyes, the most tender and sensitive organs of the body; and it is, I believe, always accompanied by a sudden and forcible inspiration, which is the natural preparation for any violent effort. But when a man or horse starts his heart beats wildly against his ribs, and here it may be truly said we have an organ which has never been under the control of the will, partaking in the general reflex movements of the body.

The contraction of the iris when the retina is stimulated by a bright light is another instance of a movement, which it appears cannot possibly have been at first voluntarily performed and then fixed by habit; for the iris is not known to be under the conscious control of the will in any animal.[1] The radiation of nerve-force from strongly-excited nerve-cells to other connected cells, as in the case of a bright light on the retina causing a sneeze, may aid us in understanding how some reflex actions originated. A radiation of nerve-force of this kind, if it caused a movement tending to lessen the primary irritation, as in the case of the contraction of the iris preventing too much light from falling on the retina, might afterwards have been taken advantage of and modified for this special purpose.

Reflex actions are in all probability liable to slight variations, as are all corporeal structures and instincts; and any variations which were beneficial would tend to be preserved and inherited. Thus reflex actions, when once gained for one purpose, might afterwards be modified so as to serve for some distinct purpose.

Associated Habitual Movements in the Lower Animals.—Dogs, when they wish to go to sleep on a

[1] Professor Beer, of Bonn, is said by Lewes (*Physical Basis of Mind*) to have had the power of contracting or dilating the pupils at will. "Here ideas act as motors. When he thinks of a very dark space the pupil dilates, when of a very bright spot the pupil contracts."

carpet or other hard surface, generally turn round and round and scratch the ground with their fore-paws in a senseless manner, as if they intended to trample down the grass and scoop out a hollow, as no doubt their wild parents did when they lived on open grassy plains or in the woods.[1] Jackals, fennecs, and other allied animals in the Zoological Gardens treat their straw in this manner; but the keepers, after observing for some months, have never seen the wolves thus behave.

Many carnivorous animals, as they crawl towards their prey and prepare to rush or spring on it, lower their heads and crouch, partly, as it would appear, to hide themselves, and partly to get ready for their rush; and this habit in an exaggerated form has become hereditary in our pointers and setters. Now I have noticed scores of times that when two strange dogs meet on an open road, the one which first sees the other, though at the distance of one or two hundred yards, after the first glance always lowers his head, generally crouches a little, or even lies down; that is, he takes the proper attitude for concealing himself and for making a rush or spring, although the road is quite open and the distance great. Again, dogs of all kinds, when intently watching and slowly approaching their prey, frequently keep one of their fore-legs doubled up for a long time, ready for the next cautious step; and this is eminently characteristic of the pointer. They behave in the same manner whenever their attention is aroused. I have seen a dog at the foot of a high wall, listening attentively to a sound on the opposite side, with one leg doubled up; and in this case there could have been no intention of making a cautious approach.

Dogs after voiding their excrement often make with

[1] It appears that Esquimaux dogs never turn round before lying down, and this fact is in harmony with the above explanation, for the Esquimaux dogs cannot, for countless generations, have had an opportunity of trampling for themselves a sleeping-place in grass.

all four feet a few scratches backwards, even on a bare stone pavement, as if for the purpose of covering up their excrement with earth, in nearly the same manner as do cats. Wolves and jackals behave in the Zoological Gardens in exactly the same manner, yet, as I am assured by the keepers, neither wolves, jackals, nor foxes, when they have the means of doing so, ever cover up their excrement, any more than do dogs. Hence, if we rightly understand the meaning of the above cat-like habit, we have a purposeless remnant of an habitual movement, which was originally followed by remote progenitors of the dog-genus for a definite purpose, and which has been retained for a prodigious length of time.

Dogs scratch themselves by a rapid movement of one of their hind feet; and when their backs are rubbed with a stick so strong is the habit that they cannot help rapidly scratching the air or the ground in a useless and ludicrous manner.

Horses scratch themselves by nibbling those parts of their bodies which they can reach with their teeth; but more commonly one horse shows another where he wants to be scratched, and they then nibble each other. A friend whose attention I had called to the subject observed that, when he rubbed his horse's neck, the animal protruded his head, uncovered his teeth, and moved his jaws, exactly as if nibbling another horse's neck, for he could never have nibbled his own neck. If a horse is much tickled, as when curry-combed, his wish to bite something becomes so intolerably strong that he will clatter his teeth together, and, though not vicious, bite his groom. At the same time, from habit he closely depresses his ears, so as to protect them from being bitten, as if he were fighting with another horse.

A horse when eager to start on a journey makes the nearest approach which he can to the habitual movement of progression by pawing the ground. Now, when horses in their stalls are about to be fed and are eager for their corn, they paw the pavement or the straw. Two of my horses thus behave when they see or hear the corn given to their neighbours.

Kittens, puppies, young pigs, and probably many other young animals, alternately push with their fore-feet against the mammary glands of their mothers, to excite a freer secretion of milk. Now it is very common with young cats, and not at all rare with old ones, when comfortably lying on a soft substance, to pound it alternately with their fore-feet; their toes being spread out and claws slightly protruded, precisely as when sucking their mother.

The wonderful power which a chicken possesses, only a few hours after being hatched, of picking up small particles of food, seems to be started into action through the sense of hearing; for with chickens hatched by artificial heat a good observer found that making a noise in imitation of the hen-mother first taught them to peck at their meat.

The Sheldrake feeds on the sands left uncovered by the tide; and when a wormcast is discovered "it begins patting the ground with its feet, dancing, as it were, over the hole"; and this makes the worm come to the surface. Now Mr. St. John says that when his tame Sheldrakes "came to ask for food they patted the ground in an impatient and rapid manner." This, therefore, may almost be considered as their expression of hunger. Mr. Bartlett informs me that the Flamingo and the Kagu, when anxious to be fed, beat the ground with their feet in the same odd manner. So again Kingfishers, when they catch a fish, sometimes[1] beat it until it is killed; and in the Zoological Gardens they occasionally beat the raw meat, with which they are fed, before devouring it.

[1] It is not correct to say that kingfishers always act in this manner. See Mr. C. C. Abbott, *Nature*, March 13, 1873, and Jan. 21, 1875.

CHAPTER II

GENERAL PRINCIPLES OF EXPRESSION—*continued*

WHEN a dog approaches a strange dog or man in a savage or hostile frame of mind he walks upright and very stiffly; his head is slightly raised, or not

FIG. 1.—Dog approaching with hostile intentions. Note *advanced* ears and erected hairs. Compare same dog in Fig. 2.

much lowered; the tail is held erect and quite rigid; the hairs bristle, especially along the neck and back; the pricked ears are directed forwards, and the eyes have a fixed stare (Fig. 1). These actions follow from the dog's intention to attack his enemy, and are thus to a large extent intelligible. As he prepares to spring with a savage growl on his enemy the canine teeth are uncovered, and the ears are pressed close backwards on the head. Let us now suppose that the dog suddenly discovers that the man whom he is approaching is not a stranger, but his master; and let it be observed how completely

FIG. 2.—Dog in a humble and affectionate frame of mind. Note the *drooping* ears and tail, depressed hairs, and the "supplicating" attitude generally. Compare same dog in Fig. 1.

and instantaneously his whole bearing is reversed. Instead of walking upright, the body sinks downwards or even crouches, and is thrown into flexuous movements; his tail, instead of being held stiff and upright, is lowered and wagged from side to side; his hair instantly becomes smooth; his ears are depressed and drawn backwards, but not closely to the head; and his lips hang loosely (Fig. 2). From the drawing back of the ears the eyelids become elongated, and the eyes no longer appear round and staring. Not one of the above movements, so clearly expressive of affection, is of the least direct service to the animal. They are explicable, as far as I can see, solely from being in complete opposition to the attitude and movements which are assumed when a dog intends to fight, and which consequently are expressive of anger. It is, however, not a little difficult to represent affection in a dog, as the essence of the expression lies in the continuous flexuous movements of his tail.

When a cat is threatened by a dog it arches its back, erects its hair, opens its mouth and spits (Fig. 6). This well-known attitude is expressive of terror combined with anger. Rage or anger may be observed when two cats are fighting together; and I have seen it well exhibited by a savage cat whilst plagued by a boy. The attitude is almost exactly the same as that of a tiger disturbed and growling over its food. The animal assumes a crouching position, with the body extended; and the whole tail, or the tip alone, is lashed or curled from side to side. The hair is not erect. Thus far, the attitude and movements are nearly the same as when the animal is prepared to spring on its prey, and when, no doubt, it feels savage. But when preparing to fight there is this difference, that the ears are closely pressed backwards; the mouth is partially opened, showing the teeth; the forefeet are occasionally struck out with protruded claws; and the animal occasionally utters a fierce growl (Fig. 3).

Now look at a cat in a directly opposite frame of mind, whilst feeling affectionate and caressing her master; and mark how opposite is her attitude in every respect. She now stands upright with her back slightly arched, which makes the hair appear rather rough, but it does not bristle; her tail, instead of being extended and lashed from side to side, is held quite stiff and perpendicularly upwards; her ears are erect; her mouth is closed; and she rubs against her master with a purr instead of a growl (Fig. 4). Let it further be observed how widely different is

FIG. 3.—Domestic Cat, savage and prepared to fight. Observe the *retracted* ears, exposed canines, and the raised fore-leg ready to scratch. Compare with Figs. 4 and 6.

the whole bearing of an affectionate cat from that of a dog, when, with his body crouching and flexuous, his tail lowered and wagging, and ears depressed,

FIG. 4.—Domestic Cat in an affectionate frame of mind. Note the raised tail and compare the general attitude with that in Figs. 3 and 6.

he caresses his master (Fig. 2). This contrast in the attitudes and movements of these two carnivorous animals, under the same pleased and affectionate frame of mind, can be explained, as it appears to me, solely by their movements standing in complete antithesis to those which are naturally assumed when they feel savage and are prepared to fight or seize their prey.

With social animals the power of intercommunication between the members of the same community—and with other species, between the opposite sexes, as well as between the young and the old—

[*F. W. Bond.*

FIG. 5.—The Lynx (Caracal) with hate and fury boiling in his breast.

is of the highest importance to them. This is generally effected by means of the voice, but it is certain that gestures and expressions are to a certain extent mutually intelligible. Man not only uses inarticulate cries, gestures, and expressions, but, by a process, completed by innumerable steps, half-consciously made, has invented articulate language. Any one who has watched monkeys will not doubt that they perfectly understand each other's gestures and expression, and to a large extent those of man. An animal when going to attack another, or when afraid of another, often makes itself appear terrible (Fig. 5), by erecting its hair, thus increasing the apparent bulk of its body, by showing its teeth, or brandishing its horns, or by uttering fierce sounds.

FIG. 6.—Domestic Cat threatening a dog. Note especially the apparent increase in size and contrast this figure with Figs. 3 and 4.

With conventional signs which are not innate, such as those used by the deaf and dumb and by savages, the principle of antithesis has been partially brought into play. The Cistercian monks thought it sinful to speak; and, as they could not avoid holding some communication, they invented a gesture-language, in which the principle of opposition seems to have been employed. Dr. Scott, of the Exeter Deaf and Dumb Institution, writes to me that "opposites are greatly used in teaching the deaf and dumb, who have a lively sense of them."

Many signs, moreover, which plainly stand in opposition to each other, appear to have had on both sides a significant origin. This seems to hold good with the signs used by the deaf and dumb for light and darkness, for strength and weakness, etc.

With mankind the best instance of a gesture standing in direct opposition to other movements, naturally assumed under an opposite frame of mind, is that of shrugging the shoulders. This expresses impotence or an apology—something which cannot be done, or cannot be avoided. The gesture is sometimes used consciously and voluntarily, but it is extremely improbable that it was at first deliberately invented, and afterwards fixed by habit; for not only do young children sometimes shrug their shoulders under the above states of mind, but the movement is accompanied by various subordinate movements.

Dogs, when approaching a strange dog, may find it useful to show by their movements that they are friendly and do not wish to fight. When two young dogs in play are growling and biting each other's faces and legs it is obvious that they mutually understand each other's gestures and manners. There seems, indeed, some degree of instinctive knowledge in puppies and kittens that they must not use their sharp little teeth or claws too freely in their play, though this sometimes happens, and a squeal is the result; otherwise they would often injure each other's eyes. When my terrier bites my hand in play, often snarling at the same time, if he bites too hard and I say "*Gently, gently,*" he goes on biting, but answers me by a few wags of the tail, which seems to say "Never mind; it is all fun." Although dogs do thus express, and may wish to express, to other dogs and to man that they are in a friendly state of mind, it is incredible that they could ever have deliberately thought of drawing back and depressing their ears, instead of holding them erect—of lowering and wagging their tails, instead of keeping them stiff and upright, etc., because they knew that these movements stood in direct opposition to those assumed under an opposite and savage frame of mind.

Again, when a cat, or rather when some early progenitor of the species, from feeling affectionate, first slightly arched its back, held its tail perpendicularly upwards, and pricked its ears, can it be believed that the animal consciously wished thus to show that its frame of mind was directly the reverse of that when, from being ready to fight or to spring on its prey, it assumed a crouching attitude, curled its tail from side to side, and depressed its ears?

Hence for the development of the movements which come under the present head some other principle, distinct from the will and consciousness, must have intervened. This principle appears to be that every movement which we have voluntarily performed throughout our lives has required the action of certain muscles; and when we have performed a directly opposite movement an opposite set of muscles has been habitually brought into play—as in turning to the right or to the left, in pushing away or pulling an object towards us, and in lifting or lowering a weight. So strongly are our intentions and movements associated together that if we eagerly wish an object to move in any direction we can hardly avoid moving our bodies in the same direction, although we may be perfectly aware that this can have no influence. A good illustration of this fact is the grotesque movements of a young and eager billiard-player whilst watching the course of his ball. A person in a passion, telling any one to be gone, generally moves his arm as if to push him away, although the offender may not be standing near, and although there may be not the least need to explain by a gesture what is meant. On the other hand, if we eagerly desire some one to approach us closely we act as if pulling him towards us; and so in innumerable other instances.

As the performance of ordinary movements of an opposite kind, under opposite impulses of the will, has become habitual in us and in the lower animals, so when actions of one kind have become firmly associated with any sensation or emotion it appears natural that actions of a directly opposite kind, though of no use, should be unconsciously performed through habit and association, under the influence of a directly opposite sensation or emotion. On this principle alone can I understand how the gestures and expressions which come under the present head of antithesis have originated.

CHAPTER III

GENERAL PRINCIPLES OF EXPRESSION

—concluded

WHEN the sensorium is strongly excited nerve-force is generated in excess, and is transmitted in certain directions dependent on the connection of the nerve-cells, and, as far as the muscular system is concerned, on the nature of the movements which have been habitually practised. Or the supply of nerve-force may, as it appears, be interrupted. Of course every movement which we make is determined by the constitution of the nervous system.

A good case is that of the trembling of the muscles, which is common to man and to many, or most, of the lower animals. Trembling is of no service, often of much disservice, and cannot have been at first acquired through the will and then rendered habitual in association with any emotion. I am assured by an eminent authority that young children do not tremble but go into convulsions under circumstances which would induce excessive trembling in adults. Trembling is excited in different individuals in very different degrees, and by the most diversified causes—by cold to the surface, before fever-fits; in blood-poisoning, delirium tremens, and other diseases; by general failure of power in old age; by exhaustion after excessive fatigue, etc. Of all emotions fear is notoriously the most apt to induce trembling; but so do occasionally great anger and joy. I remember once seeing a boy who had just shot his first snipe on the wing, and his hands trembled to such a degree from delight that he could not for some time reload his gun;[1] and

[1] The boy in question was himself. See *Life and Letters of Charles Darwin*, Vol. I. p. 34.

I have heard of an exactly similar case with an Australian savage to whom a gun had been lent. Fine music, from the vague emotions thus excited, causes a shiver to run down the backs of some persons. There seems to be very little in common in the above several physical causes and emotions to account for trembling. As trembling is sometimes caused by rage, and as it sometimes accompanies great joy, it would appear that any strong excitement of the nervous system interrupts the steady flow of nerve-force to the muscles.

The manner in which the secretions of the alimentary canal and of certain glands—as the liver, kidneys, or mammæ—are affected by strong emotions is another excellent instance of the direct action of the sensorium on these organs, independently of the will or of any serviceable associated habit.

The heart, which goes on uninterruptedly beating night and day in so wonderful a manner, is extremely sensitive to external stimulants. The least excitement of a sensitive nerve reacts on the heart; even when a nerve of an animal under experiment is touched so slightly that no pain can possibly be felt. Hence when the mind is strongly excited we might expect that it would instantly affect in a direct manner the heart; and this is universally acknowledged and felt to be the case. When the heart is affected it reacts on the brain; and the state of the brain again reacts through the pneumo-gastric nerve on the heart; so that under any excitement there will be much mutual action and reaction between these two most important organs of the body.[1]

[1] Mosso gives an interesting account of cases in which, owing to injuries to the skull, the pulsation of the brain could be observed. He has demonstrated, by means of his plethysmograph, the effect of emotion in causing a diminution in the volume of the arm, etc.; and, by means of his balance, he has shown the flow of blood to the brain under very small stimuli—for instance, when a slight noise, not sufficient to wake the

The vaso-motor system, which regulates the diameter of the small arteries, is directly acted on by the sensorium, as we see when a man blushes from shame; but in this latter case the checked transmission of nerve-force to the vessels of the face can, I think, be partly explained through habit. We shall also be able to throw some light, though very little, on the involuntary erection of the hair under the emotions of terror and rage.

When animals suffer from an agony of pain, they generally writhe about with frightful contortions; and those which habitually use their voices utter piercing cries or groans. Almost every muscle of the body is brought into strong action. With man the mouth may be closely compressed, or more commonly the lips are retracted, with the teeth clenched or ground together. There is said to be "gnashing of teeth" in hell; and I have plainly heard the grinding of the molar teeth of a cow which was suffering acutely from inflammation of the bowels. The female hippopotamus in the Zoological Gardens, when she produced her young, suffered greatly; she incessantly walked about, or rolled on her sides, opening and closing her jaws, and clattering her teeth together. With man the eyes stare wildly as in horrified astonishment, or the brows are heavily contracted (Figs. 21 and 22). Perspiration bathes the body, and drops trickle down the face. The circulation and respiration are much affected. Hence the nostrils are generally dilated and often quiver; or the breath may be held until the blood stagnates in the purple face. If the agony be severe and prolonged, these signs all change; utter prostration follows, with fainting or convulsions.

---

patient, is made in the room in which he is asleep. Mosso considers the action of emotion on the vaso-motor system as adaptive. The violent action of the heart in terror is supposed to be of use as preparing the body generally for great exertion. In a similar manner he explains the pallor of terror.

A sensitive nerve when irritated transmits some influence to the nerve-cell whence it proceeds; and this transmits its influence, first to the corresponding nerve-cell on the opposite side of the body, and then upwards and downwards along the cerebro-spinal column to other nerve-cells, to a greater or less extent according to the strength of the excitement; so that, ultimately, the whole nervous system may be affected. This involuntary transmission of nerve-force may or may not be accompanied by consciousness.

An emotion may be very strong but will have little tendency to induce movements of any kind if it has not commonly led to voluntary action for its relief or gratification; and when movements are excited, their nature is, to a large extent, determined by those which have often and voluntarily been performed for some definite end under the same emotion. Great pain has urged all animals during endless generations to make the most violent and diversified efforts to escape from the cause of suffering. Even when a limb or other separate part of the body is hurt we often shake it, as if to shake off the cause, though this may obviously be impossible. Thus a habit of exerting with the utmost force all the muscles will have been established, whenever great suffering is experienced. As the muscles of the chest and vocal organs are habitually used, these will be particularly liable to be acted on, and loud, harsh screams or cries will be uttered. But the advantage derived from cries has here probably come into play in an important manner; for the young of most animals, when in distress or danger, call loudly to their parents for aid, as do the members of the same community for mutual aid.

Under the powerful emotion of rage the action of the heart is much accelerated. The face reddens, or becomes purple from the impeded return of the blood, or it may turn deadly pale. The respiration is laboured, the chest heaves, the dilated nostrils quiver, and the whole body often trembles. The voice is affected, the teeth are clenched or ground together, and the muscular system is commonly stimulated to violent, almost frantic action. But the gestures of a man in this state usually differ from the purposeless writhings of one in agony; for they represent more or less plainly the act of striking, or fighting with, an enemy.

Animals of all kinds, and their progenitors before them, when attacked or threatened by an enemy, have exerted their utmost powers in fighting and defending themselves. Unless an animal does thus act, or has the intention, or at least the desire, to attack its enemy, it cannot properly be said to be enraged.

The heart no doubt will likewise be affected in a direct manner; but it will also in all probability be affected through habit; and all the more so from not being under the control of the will. We know that any great exertion affects the heart through mechanical and other principles; and it has been shown that nerve-force flows readily through the habitually used nerves of motion and sensation.

A man when moderately angry, or even when enraged, may command the movements of his body, but he cannot prevent his heart beating rapidly. His chest will perhaps give a few heaves, and his nostrils just quiver, for the movements of respiration are only in part voluntary. In like manner those muscles of the face which are least obedient to the will sometimes alone betray a slight and passing emotion. The glands again are wholly independent of the will, and a man suffering from grief may command his features, but cannot always prevent the tears from coming into his eyes. A hungry man, if tempting food is placed before him, may not show his hunger by any outward gesture, but he cannot check the secretion of saliva.

Under a transport of joy or of vivid pleasure, there is a strong tendency to various purposeless movements, and to the utterance of various sounds. We see this in our children, in their loud laughter, clapping of hands, and jumping for joy; in the bounding and barking of a dog when going out to walk with his master; and in the frisking of a horse when turned out into an open field. Joy quickens the circulation, and this stimulates the brain, which again reacts on the whole body. The above purposeless movements and increased heart-action may be attributed in chief part to the excited state of the sensorium, and to the consequent undirected overflow of nerve-force. It deserves notice that it is chiefly the anticipation of a pleasure, and not its actual enjoyment, which leads to purposeless and extravagant movements of the body, and to the utterance of various sounds. We see this in our children when they expect any great pleasure or treat; and dogs, which have been bounding about at the sight of a plate of food, when they get it do not show their delight by any outward sign, not even by wagging their tails. Now, with animals of all kinds the acquirement of almost all their pleasures, with the exception of those of warmth and rest, has long been associated with active movements, as in the hunting or search for food, and in their courtship. Moreover, the mere exertion of the muscles after long rest or confinement is in itself a pleasure, as we ourselves feel, and as we see in the play of young animals. Therefore on this latter principle alone we might perhaps expect that vivid pleasure would be apt to show itself conversely in muscular movements.

When an animal is alarmed it almost always stands motionless for a moment, in order to collect its senses and to ascertain the source of danger (Fig. 7), and sometimes for the sake of escaping detection. But headlong flight soon follows, with no husbanding of the strength as in fighting, and the animal continues to fly as long as the danger lasts, until utter prostration, with failing respiration and circulation, with all the muscles quivering and profuse sweating, renders further flight impossible. Hence it does not seem improbable that the principle of associated habit may in part account for, or at least augment, some of the characteristic symptoms of extreme terror.

No emotion is stronger than maternal love; but a mother may feel the deepest love for her helpless

[*F. W. Bond.*

FIG. 7.—Wild (feral) Cat from Scotland. The attitude is one of alertness and suspicion.

infant and yet not show it by any outward sign; or only by slight caressing movements, with a gentle smile and tender eyes (Fig. 8). But let any one intentionally injure her infant, and see what a change! How she starts up with threatening aspect, how her eyes sparkle and her face reddens, how her bosom heaves, nostrils dilate, and heart beats; for anger, and not maternal love, has habitually led to action. The love between the opposite sexes is widely different from maternal love; and when lovers meet their hearts beat quickly, their breathing is hurried, and their faces flush; for this love is active, not like that of a mother for her infant.

Pain, if severe, soon induces extreme depression or prostration; but it is at first a stimulant and excites to action, as we see when we whip a horse, and as is shown by the horrid tortures inflicted in

[*Topical Press.*

FIG. 8.—Lioness and Cub. Mark the placid expression of the mother. The cub, by its retracted ears and exposed canines, indicates that it wants to play at "fighting."

foreign lands on exhausted dray-bullocks, to rouse them to renewed exertion. Prolonged pain causes a marked and lasting lowering of the temperature. Fear again is the most depressing of all the emotions, and, as in the case of pain, lowers the temperature; and it soon induces utter, helpless prostration, as if in consequence of, or in association with, the most violent and prolonged attempts to escape from the danger, though no such attempts have actually been made. Nevertheless, even extreme fear often acts at first as a powerful stimulant. A man or animal driven through terror to desperation is endowed with wonderful strength, and is notoriously dangerous in the highest degree.

Chapter IV

MEANS OF EXPRESSION IN ANIMALS

The Emission of Sounds.—With many kinds of animals, man included, the vocal organs are efficient in the highest degree as a means of expression. We have seen that when the sensorium is strongly excited the muscles of the body are generally thrown into violent action; and, as a consequence, loud sounds are uttered, however silent the animal may generally be, and although the sounds may be of no use. Hares and rabbits, for instance, never, I believe, use their vocal organs except in the extremity of suffering,[1] as when a hare is wounded by the sportsman, or when a rabbit is caught by a stoat. Cattle and horses suffer great pain in silence; but when this is excessive, and especially when associated with terror, they utter fearful sounds. I have often recognized, from a distance on the Pampas, the agonized death-bellow of the cattle when caught by the lasso and hamstrung. It is said that horses, when attacked by wolves, utter loud and peculiar screams of distress.[2]

Involuntary and purposeless contractions of the muscles of the chest and glottis, excited in the above manner, may have first given rise to the emission of vocal sounds. But the voice is now largely used by

[1] Mr. J. B. Dunbar, in a letter to Darwin, states that hares cry to their young, and that the cry can be called forth by removing the young hare from where the mother left it. The cry is said to be quite different from the scream of the wounded hare.

[2] A lady communicated the following description of a horse screaming: "In a crowd in London the horse fell and got under the wheel of a carriage; the scream rang in our ears for days after, as the most expressive of agony that we had ever heard."

many animals for various purposes; and habit seems to have played an important part in its employment under other circumstances. Social animals, from habitually using their vocal organs as a means of intercommunication, use them on other occasions much more freely than other animals. But there are marked exceptions to this rule—for instance, with the rabbit. The principle, also, of association, which is so widely extended in its power, has likewise played its part. Hence it follows that the voice, from having been habitually employed as a serviceable aid under certain conditions, inducing pleasure, pain, rage, etc., is commonly used whenever the same sensations or emotions are excited, under quite different conditions, or in a lesser degree.

The sexes of many animals incessantly call for each other during the breeding-season; and in not a few cases the male endeavours thus to charm or excite the female. This, indeed, seems to have been the primeval use and means of development of the voice, as I have attempted to show in my *Descent of Man*. Thus the use of the vocal organs will have become associated with the anticipation of the strongest pleasure which animals are capable of feeling. Animals which live in society often call to each other when separated, and evidently feel much joy at meeting; as we see with a horse, on the return of his companion, for whom he has been neighing. The mother calls incessantly for her lost young ones; for instance, a cow for her calf; and the young of many animals call for their mothers. When a flock of sheep is scattered the ewes bleat incessantly for their lambs, and their mutual pleasure at coming together is manifest. Woe betide the man who meddles with the young of the larger and fiercer quadrupeds if they hear the cry of distress from their young. Rage leads to the violent exertion of all the muscles, including those of the voice; and some animals, when enraged, endeavour to strike terror into their enemies by its power and harshness, as the lion does by roaring and the dog by growling. I infer that their object is to strike terror, because the lion at the same time erects the hair of its mane, and the dog the hair along its back, and thus they make themselves appear as large and terrible as possible. Rival males try to excel and challenge each other by their voices, and this leads to deadly contests. Thus the use of the voice will have become associated with the emotion of anger, however it may be aroused. Intense pain, like rage, leads to violent cries, and the exertion of screaming by itself gives some relief; and thus the use of the voice will have become associated with suffering of any kind.

With the dog the bark of anger and that of joy do not differ much, though they can be distinguished. It is not probable that any precise explanation of the cause of each particular sound, under different states of the mind, will ever be given. Some animals, after being domesticated, have acquired the habit of uttering sounds which were not natural to them. Thus domestic dogs, and even tamed jackals, have learnt to bark, which is a noise not proper to any species of the genus, with the possible exception of *Canis latrans* of North America. Some breeds, also, of the domestic pigeon have learnt to coo in a new and quite peculiar manner.

Mr. Herbert Spencer has clearly shown that the character of the human voice alters much under the influence of various emotions, in loudness and in quality—that is, in resonance and *timbre*, in pitch and intervals. No one can listen to an eloquent orator or preacher, or to a man calling angrily to another, or to one expressing astonishment, without being struck with the truth of Mr. Spencer's remarks. It is curious how early in life the modulation of the voice becomes expressive. With one of my children, under the age of two years, I clearly perceived that his humph of assent was rendered by a slight modulation strongly emphatic, and that by a peculiar whine his negative expressed obstinate determination.

The habit of uttering musical sounds was first

developed, as a means of courtship, in the early progenitors of man, and thus became associated with the strongest emotions—those of ardent love, rivalry, and triumph. That animals utter musical notes is familiar to every one, as we may daily hear in the singing of birds. It is a more remarkable fact that one of the Gibbon apes produces an exact octave of musical sounds, ascending and descending the scale by half-tones; so that this ape may be said to sing. From this fact, and from the analogy of other animals, I have been led to infer that the progenitors of man probably uttered musical tones, before they had acquired the power of articulate speech; and that consequently, when the voice is used under any strong emotion, it tends to assume, through the principle of association, a musical character. With some of the lower animals the males employ their voices to please the females, and take pleasure in their own vocal utterances.

That the pitch of the voice bears some relation to certain states of feelings is tolerably clear. A person gently complaining of ill-treatment, or slightly suffering, almost always speaks in a high-pitched voice. Dogs, when a little impatient, often make a high piping note through their noses, which at once strikes us as plaintive.[1] Monkeys express astonishment by a half-piping, half-snarling noise; anger or impatience by repeating the sound *hu hu* in a deeper, grunting voice; and fright or pain by shrill screams. On the other hand, with mankind, deep groans and high-piercing screams equally express pain. Laughter may be either high or low; so that, with adult men, the sound partakes of the character of the vowels (as pronounced in German) *O* and *A*; whilst with children and women it has more of the character of *E* and *I*; and these latter vowel-sounds have a higher pitch than the former; yet both tones of laughter equally express enjoyment or amusement.

[1] Mr. Tylor, in his discussion on this subject, alludes to the whining of the dog.

We can see reasons for the association of certain kinds of sounds with certain states of mind. A scream, for instance, uttered by one of the members of a community, as a call for assistance, will naturally be loud, prolonged, and high, so as to penetrate to a distance. Helmholtz has shown [1] that, owing to the shape of the internal cavity of the human ear and its consequent power of resonance, high notes produce a particularly strong impression. When male animals utter sounds in order to please the females they would naturally employ those which are sweet to the ears of the species; and it appears that the same sounds are often pleasing to widely different animals, as we ourselves perceive in the singing of birds and even in the chirping of certain tree-frogs giving us pleasure. On the other hand, sounds produced in order to strike terror into an enemy would naturally be harsh or displeasing.

The interrupted, laughing, or tittering sounds made by man and by various kinds of monkeys when pleased are as different as possible from the prolonged screams of these animals when distressed. The deep grunt of satisfaction uttered by a pig when pleased with its food is widely different from its harsh scream of pain or terror.

When young infants cry they open their mouths widely, and this, no doubt, is necessary for pouring forth a full volume of sound; but the mouth then assumes an almost quadrangular shape, depending on the firm closing of the eyelids and consequent drawing up of the upper lip. This square shape of the mouth may possibly modify the sound, for we know from the researches of Helmholtz that the form of the cavity of the mouth determines the nature and pitch of the vowel sounds produced.

When anyone is startled or suddenly astonished there is an instantaneous tendency, likewise from an intelligible cause—namely, to be ready for prolonged

[1] Helmholtz has also fully discussed the relation of the form of the cavity of the mouth to the production of vowel-sounds.

exertion, to open the mouth widely, so as to draw a deep and rapid inspiration. When the next full expiration follows the mouth is slightly closed, and the lips are somewhat protruded; and this form of the mouth, if the voice be at all exerted, produces, according to Helmholtz, the sound of the vowel *O*. Certainly a deep sound of a prolonged *Oh!* may be heard from a whole crowd of people immediately after witnessing any astonishing spectacle. If, together with surprise, pain be felt, there is a tendency to contract all the muscles of the body, including those of the face, and the lips will then be drawn back; and this will perhaps account for the sound becoming higher and assuming the character of *Ah!* or *Ach!* As fear causes all the muscles of the body to tremble, the voice naturally becomes tremulous, and at the same time husky from the salivary glands failing to act.

Sounds produced by other than vocal means are likewise expressive. Rabbits stamp loudly on the ground as a signal to their comrades; and if a man knows how to do so properly, he may on a quiet evening hear the rabbits answering him all around. These animals, as well as some others, also stamp on the ground when made angry. Porcupines rattle their quills and vibrate their tails when angered; and one behaved in this manner when a live snake was placed in its compartment. The quills on the tail are very different from those on the body: they are short, hollow, thin like a goose-quill, with their ends open. When the tail is rapidly shaken these hollow quills strike against each other and produce a peculiar continuous sound. We can, I think, understand why this special sound-producing instrument exists in porcupines. They are nocturnal animals; and if they scented or heard a prowling beast of prey it would be a great advantage to them in the dark to give warning to their enemy what they were, and that they were furnished with dangerous spines. They would thus escape being attacked. They are so fully conscious of the power of their weapons that when enraged they will charge backwards with their spines erected.

Many birds during courtship produce diversified sounds. Storks, when excited, make a loud clattering noise with their beaks. Some snakes produce a grating or rattling noise. Many insects stridulate by rubbing together specially modified parts of their hard integuments. This stridulation generally serves as a sexual charm or call; but it is likewise used to express different emotions. Every one who has attended to bees knows that their humming changes when they are angry; and this serves as a warning that there is danger of being stung.

Erection of the Dermal Appendages.—Hardly any expressive movement is so general as the involuntary erection of the hairs, feathers, and other dermal appendages; for it is common throughout three of the great vertebrate classes.[1] These appendages are erected under the excitement of anger or terror; more especially when these emotions are combined, or quickly succeed each other. The action serves to make the animal appear larger and more frightful to its enemies or rivals, and is generally accompanied by various voluntary movements adapted for the same purpose, and by the utterance of savage sounds.

When the Chimpanzee and Orang are suddenly frightened, as by a thunderstorm, or when they are made angry, as by being teased, their hair becomes erect. I saw a chimpanzee who was alarmed at the sight of a black coalheaver, and the hair rose all over his body; he made little starts forward as if to

[1] Rev. S. J. Whitmee describes the erection of the dorsal and anal fins of fishes in anger and fear. He suggests that the erection of the spines gives protection against carnivorous fish, and if this is so it is not hard to understand the association of such movements with these emotions. Mr. F. Day criticizes Mr. Whitmee's conclusions, but the description by Mr. Whitmee of a spiny fish sticking in the throat of a bigger fish and being finally ejected seems to prove that the spines are useful.

attack the man and with the hope of frightening him. The Gorilla, when enraged, is described as having his crest of hair "erect and projecting forward, his nostrils dilated, and his under lip thrown down; at the same time uttering his characteristic yell, designed,

FIG. 9.—Hen angry and driving away a dog from her chickens. The great apparent increase of size is brought about through every feather "standing to attention."

it would seem, to terrify his antagonists."[1] I saw the hair on the Anubis baboon, when angered, bristling along the back, from the neck to the loins. I took a stuffed snake into the monkey-house, and the hair on several of the species instantly became erect, especially on their tails. Brehm states that the American *Midas œdipus* when excited erects its mane, in order to make itself as frightful as possible.

With the Carnivora the erection of the hair seems

[1] Huxley's *Evidence as to Man's Place in Nature.*

to be almost universal, often accompanied by threatening movements, the uncovering of the teeth, and the utterance of savage growls (Fig. 5). In the mongoose I have seen the hair on end over nearly the whole body, including the tail; and the dorsal crest is erected in a conspicuous manner by the Hyæna and the Aard Wolf of Africa. An enraged lion erects his mane. The bristling of the hair along the neck and back of

FIG. 10.—Enraged Swan driving away an intruder. Note the open mouth and, in imagination, the terrifying hiss being emitted.

the dog, and over the whole body of the cat, is familiar to every one. I have often noticed that the hair of a dog is particularly liable to rise if he is half angry and half afraid, as on beholding some object only indistinctly seen in the dusk.

Birds when angry ruffle their feathers and spread out their wings and tail (Figs. 9 and 10). With their plumage in this state, they rush at each other with open beaks and threatening gestures. A hybrid goldfinch of a most irascible disposition, when approached too closely by a servant, instantly assumed the appearance of a ball of ruffled feathers. Birds when frightened pull all their feathers down close to the body, and their consequently diminished size is often astonishing. As soon as they recover from their fear the first thing which they do is to shake out their feathers. The habit is intelligible in birds, since it assists them, when in danger, to escape detection.

We thus see how generally throughout the two higher vertebrate classes, and with some reptiles, the dermal appendages are erected under the influence of anger and fear. The movement is effected by the contraction of minute, unstriped, involuntary muscles, called *arrectores pili*, which are attached to the capsules of the separate hairs, feathers, etc. The erection of the hair is, however, aided in some cases by the striped and voluntary muscles of the underlying *panniculus carnosus*. It is by the action of these latter muscles that the hedgehog erects its spines. It appears that striped fibres extend from the panniculus to some of the larger hairs, such as the vibrissæ of certain quadrupeds. The *arrectores pili* contract not only under the above emotions, but also from the application of cold to the surface. I remember that my mules and dogs, brought from a lower and warmer country, after spending a night on the bleak Cordillera, had the hair all over their bodies as erect as under the greatest terror. We see the same action in our own *goose-skin* during a chill. Tickling a neighbouring part of the skin also causes the erection and protrusion of the hairs.

From these facts it is manifest that the erection of the dermal appendages is a reflex action, independent of the will. In a large number of animals, belonging to widely distinct classes, the erection of the hair or feathers is almost always accompanied by various voluntary movements—by threatening gestures, opening the mouth, uncovering the teeth (Figs. 5, 6, 11), spreading out of the wings and tail by birds (Figs. 9, 10), and by the utterance of harsh sounds; and the purpose of these voluntary movements is unmistakable. Therefore it seems hardly credible that the co-ordinated erection of the dermal appendages, by which the animal is made to appear larger and more terrible to its enemies or rivals, should be altogether an incidental and purposeless result of the disturbance of the sensorium.

We here encounter a great difficulty. How can the contraction of the unstriped and involuntary *arrectores pili* have been co-ordinated with that of various voluntary muscles for the same special purpose? If we could believe that the arrectores primordially had been voluntary muscles and had since become involuntary, the case would be comparatively simple.

We may admit that originally the *arrectores pili* were slightly acted on in a direct manner, under the influence of rage and terror, by the disturbance of the nervous system; as is undoubtedly the case with our so-called *goose-skin*. Animals have been repeatedly excited by rage and terror during many generations; and consequently the direct effects of the disturbed nervous system on the dermal appendages will almost certainly have been increased through habit and through the tendency of nerve-force to pass readily along accustomed channels. As soon as the power of erection had thus been strengthened or increased with animals, they must often have seen the hairs or feathers erected in rival and enraged males, and the bulk of their bodies thus increased (Figs. 9, 10). In this case it appears possible that they might have wished to make themselves appear larger and more terrible to their enemies, by voluntarily assuming a threatening attitude and uttering harsh cries; such attitudes and utterances after a time becoming through habit instinctive. In this manner actions performed by the contraction of voluntary muscles might have been combined for the same special purpose with those effected by involuntary muscles. We must not overlook the part which variation and natural selection may have played; for the males which succeeded in making themselves appear the most terrible to their rivals, or to their other enemies, will on an average have left more offspring to inherit their characteristic qualities than have other males.[1]

The Inflation of the Body, and other Means of Exciting Fear in an Enemy.—Certain Amphibians and Reptiles, which either have no spines to erect or no muscles by which they can be erected, enlarge themselves when alarmed or angry by inhaling air. This is well known to be the case with toads and frogs. The latter animal is made, in Æsop's fable of the "Ox and the Frog," to blow itself up from vanity and envy until it burst. This action must have been observed during the most ancient times, as the word *toad* expresses in several of the languages of Europe the habit of swelling. The primary purpose probably was to make the body appear as large and frightful as possible to an enemy; but another and secondary advantage is thus gained. When frogs are seized by snakes, which are their chief enemies, they enlarge themselves wonderfully; so that if the snake be of small size it cannot swallow the frog, which thus escapes being devoured. The males of some lizards, when fighting together during their courtship, expand their throat-pouches or frills and erect their dorsal crests. But Dr. Günther does not believe that they can erect their separate spines or scales.

Chameleons and some other lizards inflate themselves when angry. Thus a species inhabiting Oregon, the *Tapaya Douglasii*, is slow in its movements and does not bite, but has a ferocious aspect. "When irritated it springs in a most threatening manner at anything pointed at it, at the same time opening its

[1] Dr. T. Clay Shawe is inclined to doubt that the bristling of the hairs is due to the *panniculus carnosus* rather than to the *arrectores*. But the hair on a cat's tail bristles with anger or fear; and here, as Professor Macalister tells me, the effect must be due to the *arrectores*, there being no *panniculus*.

mouth wide and hissing audibly, after which it inflates its body and shows other marks of anger."

Several kinds of snakes likewise inflate themselves when irritated. The puff-adder (*Clotho arietans*) is remarkable in this respect; but I believe, after carefully watching these animals, that they do not so much act thus to make themselves appear larger as to gain a large supply of air to produce their surprisingly loud and prolonged hissing sound. The Cobras-de-capello, when irritated, enlarge themselves a little, and hiss moderately; but, at the same time, they lift their heads aloft, and dilate by means of their elongated anterior ribs the skin on each side of the neck into the so-called hood. With their widely opened mouths they then assume a terrifying aspect. An innocuous snake, the *Tropidonotus macrophthalmus* of India, likewise dilates its neck when irritated, and consequently is often mistaken for its compatriot, the deadly Cobra. This resemblance perhaps serves as some protection to the Tropidonotus. Another innocuous species, the Dasypeltis of South Africa, blows itself out, distends its neck, hisses and darts at an intruder. Many other snakes hiss under similar circumstances. They also rapidly vibrate their protruded tongues; and this may aid in increasing their terrifying appearance.

Snakes possess other means of producing sounds besides hissing. Many years ago I observed in South America that a venomous Trigonocephalus,[1] when disturbed, rapidly vibrated the end of its tail, which, striking against the dry grass and twigs, produced a rattling noise that could be distinctly heard at the distance of six feet.[2] The deadly and fierce *Echis carinata*[3] of India produces "a curious prolonged, almost hissing sound" in a very "different manner—

[1] One of the water-vipers.—C. M. B.

[2] *Journal of Researches during the Voyage of the "Beagle,"* 1843, p. 96. I here compared the rattling thus produced with that of the Rattlesnake.

[3] The sand-viper.—C. M. B.

namely, by rubbing the sides of the folds of its body against each other," whilst the head remains in almost the same position. The scales on the sides, and not on other parts of the body, are strongly keeled, with the keels toothed like a saw; and as the coiled-up animal rubs its sides together these grate against each other. Lastly, we have the well-known noise of the rattlesnake. Professor Shaler states that it is indistinguishable from that made by the male of a large cicada (an Homopterous insect), which inhabits the same district. I cannot follow Professor Shaler in believing that the rattle has been developed, by the aid of natural selection, for the sake of producing sounds which deceive and attract birds, so that they may serve as prey to the snake. I do not, however, doubt that the sounds may occasionally subserve this end. But that the rattling serves as a warning to would-be devourers appears to me much more probable. If this snake had acquired its rattle and the habit of rattling, for the sake of attracting prey, it does not seem probable that it would have invariably used its instrument when angered or disturbed. In the Zoological Gardens, when the rattlesnakes and puff-adders were greatly excited at the same time, I was much struck at the similarity of the sound produced by them; I conclude, from the threatening gestures made at the same time by snakes, that their hissing, rattling of the tail, grating of the scales and dilatation of the hood, and puffing out of the head, all subserve the same end—namely, to make them appear terrible to their enemies.[1]

It seems at first a probable conclusion that venomous snakes, such as the foregoing, from being already so well defended by their poison-fangs, would never be attacked by any enemy, and consequently would

[1] From accounts collected of the habits of the snakes of South Africa, and of the rattlesnake in North America, it does not seem improbable that the terrifying appearance of snakes and the sounds produced by them may likewise serve in procuring prey, by paralysing, or, as it is sometimes called, fascinating, the smaller animals.

have no need to excite additional terror. But this is far from being the case, for they are largely preyed on in all quarters of the world by many animals. Pigs are employed in the United States to clear districts infested with rattlesnakes, which they do most effectually.[1] Our hedgehog attacks and devours the viper. In India several kinds of hawks, and at least one mammal, the mongoose, kill cobras and other venomous species;[2] and so it is in South Africa. Therefore it is by no means improbable that any sounds or signs by which the venomous species could instantly make themselves recognized as dangerous would be of more service to them than to the innocuous species which would not be able, if attacked, to inflict any real injury.

The Means by which the Rattle of the Rattlesnake was Probably Developed.—Various animals, including lizards and snakes, either curl or vibrate their tails when excited. *Coronella Sayi*, an innocuous snake, vibrates its tail so rapidly that it becomes almost invisible; Trigonocephalus has the same habit. That the rattle has been specially developed to serve as an efficient sound-producing instrument there can hardly be a doubt; for even the vertebræ included within the extremity of the tail have been altered in shape and cohere. But there is no greater improbability in various structures, such as the rattle of the rattlesnake, the lateral scales of the sand-viper, the neck with the included ribs of the cobra, and the whole body of the puff-adder, having been modified for the sake of warning and frightening away their enemies, than in the Secretary-hawk[3] having had its whole frame

[1] Dr. R. Brown says that as soon as a pig sees a snake it rushes upon it; and a snake makes off immediately on the appearance of a pig.

[2] Dr. Günther remarks on the destruction of cobras by the mongoose, and, whilst the cobras are young, by the jungle-fowl. The peacock also eagerly kills snakes.

[3] The Secretary-bird is now generally recognized as belonging to the order *Falconiformes* containing the eagles, hawks, etc.—C. M. B.

modified for the sake of killing snakes with impunity. It is highly probable, judging from what we have before seen, that this bird would ruffle its feathers whenever it attacked a snake; and it is certain that the mongoose, when it eagerly rushes to attack a snake, erects the hair all over its body, and especially that on its tail. Some porcupines, when alarmed at the sight of a snake, rapidly vibrate their tails, thus producing a peculiar sound by the striking together of the hollow quills. So that here both the attackers and the attacked endeavour to make themselves as dreadful as possible to each other; and both possess for this purpose specialized means, which, oddly enough, are nearly the same in some of these cases. Finally, we can see that if, on the one hand, those individual snakes which were best able to frighten away their enemies escaped most from being devoured; and if, on the other hand, those individuals of the attacking enemy which were best fitted for the dangerous task of killing and devouring venomous snakes survived in larger numbers; then in the one case as in the other, beneficial variations, supposing the characters in question to vary, would commonly have been preserved through the survival of the fittest.

The Drawing Back and Pressure of the Ears to the Head.—The ears through their movements are highly expressive in many animals; but in some, such as man, the higher apes, and many ruminants, they fail in this respect. A slight difference in position serves to express in the plainest manner a different state of mind, as we may daily see in the dog; but we are here concerned only with the ears being drawn closely backwards and pressed to the head. A savage frame of mind is thus shown, but only in the case of those animals which fight with their teeth; and the care which they take to prevent their ears being seized by their antagonists accounts for this position (Figs. 3, 5, 6, 11). Consequently, through habit and association, whenever they feel slightly savage or pretend in their play to be savage, their ears are drawn back. That this is the true explanation may be inferred from the relation which exists in very many animals between their manner of fighting and the retraction of their ears.

All the Carnivora fight with their canine teeth, and all, as far as I have observed, draw their ears back when feeling savage. This may be continually seen with dogs when fighting in earnest, and with puppies

FIG. 11.—Head of snarling Dog. The retracted upper lip and ears make a gesture distinctively antagonistic.

fighting in play. The movement is different from the falling down and slight drawing back of the ears when a dog feels pleased and is caressed by his master. The retraction of the ears may likewise be seen in kittens fighting together in their play, and in full-grown cats when really savage, as before illustrated in Fig. 3. Although their ears are thus to a large extent protected, yet they often get much torn in old male cats during their mutual battles. The same movement is very striking in tigers, leopards, etc. whilst growling over their food. The lynx has remarkably long ears; and their retraction when one of these animals is approached is eminently expressive of its savage disposition (Fig. 5). Even one of the Eared Seals, *Otaria pusilla*, which has very small ears, draws them backwards when it makes a savage rush at the legs of its keeper.

When horses fight together they use their incisors for biting and their fore-legs for striking, much more than they do their hind-legs for kicking backwards. This has been observed when stallions have broken loose and have fought together. Every one recognizes the vicious appearance which the drawing back of the ears gives to a horse. This movement is very different from that of listening to a sound behind. If an ill-tempered horse in a stall is inclined to kick backwards, his ears are retracted from habit, though he has no power to bite. But when a horse throws up both hind-legs in play, as when entering an open field or when just touched by the whip, he does not generally depress his ears, for he does not then feel vicious. Guanacoes frequently fight savagely with their teeth, and the hides of several which I shot in Patagonia were deeply scored. So do camels; and both these animals, when savage, draw their ears closely backwards. Guanacoes, when not intending to bite, but merely to spit their offensive saliva from a distance at an intruder, retract their ears. Even the hippopotamus, when threatening with its widely open enormous mouth a comrade, draws back its small ears, just like a horse.

Now what a contrast is presented between the foregoing animals and cattle, sheep, or goats, which never use their teeth in fighting, and never draw back their ears when enraged![1] Although sheep and goats appear such placid animals, the males often

[1] The following note is in Darwin's handwriting, and seems to be from an early note-book :—

"The giraffe kicks with its front legs, and knocks with the back of its head, yet never puts down its ears. Good to contrast with horses."

join in furious contests. Major Ross King says of the Moose-deer in Canada:—"When two males chance to meet, laying back their ears and gnashing their teeth together, they rush at each other with appalling fury." Mr. Bartlett informs me that some species of deer fight savagely with their teeth, so that the drawing back of the ears by the moose accords with our rule. Several kinds of kangaroos fight by scratching with their fore-feet and by kicking with their hind-legs; but they never bite each other, and the keepers have never seen them draw back their ears when angered. Rabbits fight chiefly by kicking and scratching, but they likewise bite each other; and I have known one to bite off half the tail of its antagonist. At the commencement of their battles they lay back their ears, but afterwards, as they bound over and kick each other, they keep their ears erect, or move them much about.

Mr. Bartlett watched a wild boar quarrelling rather savagely with his sow; and both had their mouths open and their ears drawn backwards. But this does not appear to be a common action with domestic pigs when quarrelling. Boars fight together by striking upwards with their tusks; and Mr. Bartlett doubts whether they then draw back their ears. Elephants, which in like manner fight with their tusks, do not retract their ears, but, on the contrary, erect them when rushing at each other or at an enemy.

Rhinoceroses fight with their nasal horns, and have never been seen to attempt biting each other except in play; the keepers in the Zoological Gardens are convinced that they do not draw back their ears, like horses and dogs, when feeling savage.

Some kinds of monkeys which have movable ears, and which fight with their teeth—for instance, *Cercopithecus ruber*—draw back their ears when irritated just like dogs; and they then have a very spiteful appearance. Other kinds—and this is a great anomaly in comparison with most other animals—retract their ears, show their teeth and jabber,

[*F. W. Bond*.

FIG. 12.—A Baboon (*Chacma* or *Cynocephalus*) showing off his dental armature. The gesture is not necessarily one of rage in the apes, being often exhibited when they are tickled or otherwise pleased. It is rather a gentle reminder that the teeth are in good order, for apes that have lost their canines no longer make it.

when they are pleased by being caressed. I observed this in two or three species of Macacus, and in the *Cynopithecus niger*. This expression, owing to our familiarity with dogs, would never be recognized as one of joy or pleasure by those unacquainted with monkeys. (Fig. 12.)

Erection of the Ears.—This movement requires hardly any notice. All animals which have the power of freely moving their ears, when they are startled, or when they closely observe any object, direct their ears to the point towards which they are looking, in order to hear any sound from this quarter (Figs. 1, 7). At the same time they generally raise their heads, as their organs of sight are there situated, and some of the smaller animals rise on their hind-legs. Even those kinds which squat on the ground, or instantly flee to avoid danger, generally act momentarily in this manner, in order to ascertain the source and nature of the danger. The raised head, erected ears, and eyes directed forwards give an unmistakable expression of close attention to any animal.

Chapter V

SPECIAL EXPRESSIONS OF ANIMALS

Dogs.—So familiar is the appearance of a dog with hostile intentions—namely, with erected ears, eyes intently directed forwards, hair bristling, gait stiff, with the tail upright and rigid, that an angry man is sometimes said "to have his back up." When a tiger or wolf is suddenly roused to ferocity every muscle is in tension. This tension of the muscles and consequent stiff gait may be accounted for on the principle of associated habit, for anger has continually led to fierce struggles, and consequently to all the muscles of the body having been violently exerted.

A dog in cheerful spirits, and trotting before his master with high, elastic steps, generally carries his tail aloft, though it is not held nearly so stiffly as when he is angered. A horse, when first turned out into an open field, may be seen to trot with long elastic strides, the head and tail being held high aloft. Even cows, when they frisk about from pleasure, throw up their tails in a ridiculous fashion. So it is with various other animals. The position of the tail, however, is in part determined by special circumstances; thus, as soon as a horse breaks into a gallop he always lowers his tail, so that as little resistance as possible may be offered to the air.[1]

When a dog is on the point of springing on his antagonist he utters a savage growl; the ears are pressed closely backwards, and the upper lip (Fig. 11)

[1] Mr. Wallace suggests a different explanation. "As the whole available nervous energy is being expended in locomotion, all special muscular contractions not aiding the motion cease."

is retracted out of the way of his teeth, especially of his canines. If a dog only snarls at another, the lip is generally retracted on one side alone—namely, towards his enemy.

The movements of a dog whilst exhibiting affection (Fig. 2) stand, as we have seen, in complete antithesis to those naturally assumed under a directly opposite state of mind. Dogs also exhibit their affection by rubbing against their masters, and by showing desire to be rubbed or patted by them.

Dogs have yet another way of exhibiting their affection—namely, by licking the hands or faces of their masters. They sometimes lick other dogs, and then it is always their chops. I have also seen dogs licking cats with whom they were friends. This habit probably originated in the females carefully licking their puppies—the dearest object of their love—for the sake of cleansing them. They also often give their puppies, after a short absence, a few cursory licks, apparently from affection. Thus the habit will have become associated with the emotion of love, however it may afterwards be aroused. It is now so firmly inherited or innate that it is transmitted equally to both sexes. A female terrier of mine had her puppies destroyed, and, though at all times a very affectionate creature, she then tried to satisfy her instinctive maternal love by expending it on me; and her desire to lick my hands rose to an insatiable passion.[1]

A pleasurable and excited state of mind, associated with affection, is exhibited by some dogs in a very peculiar manner—namely, by grinning.[2]

[1] M. Baudry quotes a passage in the *Rāmāyana* where a mother, finding the dead body of her son, licks the face of the corpse with her tongue while moaning like a cow deprived of her calf.

[2] A correspondent states that in cattle the uncovering of the teeth is connected with the sexual instinct. "I was buying a bull and wished to examine his teeth, but this he would by no means allow; the natives suggested that a cow should be brought," when "the bull immediately stretched out his

Sir W. Scott's famous Scotch greyhound, Maida, had this habit, and it is common with terriers and other dogs. The upper lip during the act of grinning is retracted, as in snarling, so that the canines are exposed, and the ears are drawn backwards; but the general appearance of the animal clearly shows that anger is not felt. Sir C. Bell remarks, "Dogs, in their expression of fondness, have a slight eversion of the lips, and grin and sniff amidst their gambols, in a way that resembles laughter."

Under the expectation of any great pleasure, dogs bound and jump about in an extravagant manner and bark for joy. The tendency to bark under this state of mind is inherited; greyhounds rarely bark, whilst the Spitz-dog barks so incessantly on starting for a walk with his master that he becomes a nuisance.

Pain is expressed by dogs in nearly the same way as by many other animals, namely, by howling, writhing, and contortions of the whole body.

Attention is shown by the head being raised, with the ears erected, and eyes intently directed towards the object or quarter under observation. If it be a sound and the source is not known, the head is often turned obliquely from side to side in a most significant manner, apparently in order to judge with more exactness from what point the sound proceeds. When their attention is in any way aroused, whilst watching some object, or attending to some sound, dogs often lift up one paw and keep it doubled up, as if to make a slow and stealthy approach.

A dog under extreme terror will throw himself

---

neck and opened his lips so as to expose his teeth." He states that the practice of bringing a cow to make a bull show his teeth is a common one in India.

> "And with a courtly grin the fawning hound
> Salutes thee cow'ring, his wide op'ning nose
> Upward he curls, and his large sloe-black eyes
> Melt in soft blandishments, and humble joy."
>
> (Somerville, *The Chase*.)

down, howl, and void his excretions; but the hair does not become erect unless some anger is felt. I have seen a dog much terrified at a band of musicians who were playing loudly outside the house, with every muscle of his body trembling, with his heart palpitating so quickly that the beats could hardly be counted, and panting for breath with widely open mouth, in the same manner as a terrified man does.

Even a very slight degree of fear is invariably shown by the tail being tucked in between the legs.[1] It appears probable that the tucking in of the tail is not so much an attempt to protect it as a part of a general attempt to make the surface exposed as small as possible (*cf.* the kneeling of hyænas described below). It would thus be analogous to the shrugging of the shoulders, if M. Baudry is right in connecting this gesture with an effort to tuck in the head. This tucking in of the tail is accompanied by the ears being drawn backwards; but they are not pressed closely to the head, as in snarling, and they are not lowered, as when a dog is pleased or affectionate. When two young dogs chase each other in play, the one that runs away always keeps his tail tucked inwards. So it is when a dog, in the highest spirits, careers like a mad creature round and round his master in circles, or in figures of eight. He then acts as if another dog were chasing him. This curious kind of play, which must be familiar to every one who has attended to dogs, is particularly apt to occur after the animal has been a little startled, as by his master suddenly jumping out on him in the dusk. It appears that a dog, if chased or in danger of being struck behind or of anything falling on him, tries to withdraw as quickly as possible his whole hind-quarters, and hence the tail is drawn closely inwards.

[1] In a cuneiform inscription, nearly 5000 years old, giving an account of the Deluge, there is a description of the terror of the gods at the tempest. And "the gods like dogs with tails hidden couched down."

When two hyænas fight together, they are mutually conscious of the wonderful power of each other's jaws, and are extremely cautious. They well know that if one of their legs was seized, the bone would instantly be crushed; hence they approach each other kneeling, with their legs turned as much as possible inwards, and with their whole bodies bowed, so as not to present any salient point; the tail at the same time being closely tucked in between the legs. In this attitude they approach each other sideways, or even partly backwards. So again with deer, several of the species, when savage and fighting, tuck in their tails. When one horse in a field tries to bite the hindquarters of another in play, the tail is drawn in. On the other hand, when an animal trots with high elastic steps, the tail is almost always carried aloft.

When a dog is chased and runs away, he keeps his ears directed backwards, but still open; and this is clearly done for the sake of hearing his pursuer. From habit the ears are often held in this same position, and the tail tucked in, when the danger is obviously in front. A timid terrier of mine, when afraid of some object in front the nature of which she perfectly knows, will for a long time hold her ears and tail in this position, looking the image of discomfort. One day I went out of doors, just at the time when this dog knew that her dinner would be brought. I did not call her, but she wished much to accompany me, and at the same time she wished much for her dinner; and there she stood, first looking one way and then the other, with her tail tucked in and ears drawn back, presenting an unmistakable appearance of perplexed discomfort.

Almost all the expressive movements now described, with the exception of the grinning from joy, are innate or instinctive, for they are common to all the individuals, young and old, of all the breeds. Most of them are likewise common to the aboriginal parents of the dog—namely, the wolf and jackal;

and some of them to other species of the same group.[1] Tamed wolves and jackals, when caressed by their masters, jump about for joy, wag their tails, lower their ears, lick their master's hands, crouch down, and even throw themselves on the ground belly upwards. Wolves and jackals, when frightened, certainly tuck in their tails; and a tamed jackal has been described as careering round his master in circles and figures of eight, like a dog, with his tail between his legs.

I believe that foxes seldom lick the hands of their masters,[2] and I have been assured that when frightened they never tuck in their tails. It would thus appear that wolves, jackals, and even foxes, which have never been domesticated have acquired, through the principle of antithesis, certain expressive gestures.

Cats.—A cat when feeling savage assumes a crouching attitude and occasionally protrudes her fore-feet, with her claws extruded ready for striking. The tail is extended and lashed from side to side. The ears are drawn closely backwards and the teeth are shown. Low savage growls are uttered. We can understand why the attitude assumed by a cat when preparing to fight with another cat (Fig. 3), or in any way greatly irritated (Fig. 6), is so widely different from that of a dog approaching another dog with hostile intentions; for the cat uses her fore-feet for striking, and this renders a crouching position necessary. She is also much more accustomed than a dog to lie concealed and suddenly spring on her prey. No cause can be assigned with certainty for the tail being lashed from side to side. This habit is common to other felines—for instance, to the puma, when prepared to spring. Some kinds of lizards and snakes, too,

[1] Mr. Arthur Nicols states of a pure-bred Dingo which had been reared from a litter of wild pups, that he never saw it wag its tail or erect it when approaching a strange dog.

[2] Mr. R. M. Lloyd describes a tame fox licking the hands and face of its master.

when excited, rapidly vibrate the tips of their tails. It would appear as if, under strong excitement, there existed an uncontrollable desire for movement of some kind, and that as the tail is left free, and as its movement does not disturb the general position of the body, it is lashed about.[1]

All the movements of a cat when feeling affectionate are in complete antithesis to those just described. She now stands upright, with slightly arched back, tail perpendicularly raised, and ears erected; and she rubs her cheeks and flanks against her master or mistress (Fig. 4). The desire to rub something is so strong in cats under this state of mind, that they may often be seen rubbing themselves against the legs of chairs or tables, or against door-posts. This manner of expressing affection probably originated through association, as in the case of dogs, from the mother nursing and fondling her young; and perhaps from the young themselves loving each other and playing together. Young and even old cats when pleased alternately protrude their fore-feet with separated toes, as if pushing against and sucking their mother's teats. This habit is so far analogous to that of rubbing against something, that both apparently are derived from actions performed during the nursing period.

Cats when terrified stand at full height and arch their backs in a well-known and ridiculous fashion

[1] Does not a more utilitarian motive exist for such tail-movements than that suggested by Darwin? Both leopard and cat when preparing to spring on their prey often move the *extreme tip only* of the tail from side to side. The wriggling tail-tip rivets the attention of the prey and so facilitates its capture.

The lizard readily parts with its tail by a process of autotomy or self-amputation which takes place through a non-osseous partition in the middle of one of the caudal vertebrae. That the act is reflex and independent of volition is proved by the fact that a beheaded lizard throws off its tail when it is suddenly pinched. There is obvious survival-value in a principle that sacrifices a subordinate part to an enemy while the rest of the animal escapes.—C. M. B.

(Fig. 6). They spit, hiss, or growl. The hair over the whole body, and especially on the tail, becomes erect. The ears are drawn back, and the teeth exposed. When two kittens are playing together the one often thus tries to frighten the other. All these points of expression are intelligible; as birds ruffle their feathers and spread out their wings and tail to make themselves look as big as possible (Figs. 9, 10), so cats stand at their full height, arch their backs, and erect their hair, for the same purpose (Fig. 6). The lynx, when attacked, is said also to arch its back, but this action in tigers and lions has not been observed, and this is probably because these great felines have little cause to fear other animals.

Cats use their voices much as a means of expression, and they utter, under various emotions and desires, at least six or seven different sounds. Their purr of satisfaction is made during both inspiration and expiration. The puma, cheetah, and ocelot likewise purr; but the tiger, when pleased, emits a peculiar short snuffle, accompanied by the closure of the eyelids. It is said that the lion, jaguar, and leopard, do not purr.

Horses.—Horses when savage draw their ears back, thrust forward their heads, and partially uncover their front teeth. When inclined to kick behind, they draw back their ears and turn their eyes backwards. When pleased, as when some food is brought to them, they raise their heads, prick their ears, and looking towards their friend, often whinny. Impatience is expressed by pawing the ground.

The actions of a horse when much startled are highly expressive. One day my horse was much frightened at a drilling machine, covered by a tarpaulin, and lying on an open field. He raised his head so high, that his neck became almost perpendicular. His eyes and ears were directed intently forwards, and I could feel through the saddle the palpitations of his heart. With red dilated nostrils he snorted violently and, whirling round, would have dashed off at full speed had I not prevented him. The distension of the nostrils is not for the sake of scenting the source of danger, for when a horse smells carefully at any object and is not alarmed, he does not dilate his nostrils. A horse when panting does not breathe through his open mouth, but through his nostrils; and these consequently have become endowed with great powers of expansion. This expansion of the nostrils as well as the snorting, and the palpitations of the heart, are actions which have become firmly associated during a long series of generations with the emotion of terror; for terror has habitually led the horse to the most violent exertion in dashing away at full speed from the cause of danger.

Ruminants.—Cattle and sheep display in but slight degree their emotions. Sheep will strike viciously with their fore-feet at a dog. It seems doubtful, however, whether this action can have given rise to the stamping habit. This may be a signal, and be understood as such from its resemblance to the sound of the frightened scamper of a startled sheep. An enraged bull holds his head lowered with distended nostrils, and bellows. He also often paws the ground; but this pawing seems different from that of an impatient horse, for when the soil is loose, he throws up clouds of dust. I believe that bulls act in this manner when irritated by flies for the sake of driving them away. The wilder breeds of sheep and the chamois when startled stamp on the ground, and whistle through their noses; and this serves as a danger-signal to their comrades. The musk-ox of the Arctic regions, when encountered, likewise stamps on the ground.

Some species of deer, when savage, draw back their ears, grind their teeth, erect their hair, squeal, stamp on the ground, and brandish their horns. One day in the Zoological Gardens, a Formosan deer approached me with his muzzle raised so that the horns were pressed back on his neck. From his expression I felt sure that he was savage; he approached slowly, and as soon as he came close to the bars, he suddenly bent his head inwards, and struck his horns with great force against the railings.

Monkeys.—Monkeys express their feelings in many different ways, a fact that bears on the question whether the so-called races of man should be ranked as distinct species or varieties; for, as we shall see, the different races of man express their emotions and sensations with remarkable uniformity throughout the world. Some of the expressive actions of monkeys are closely analogous to those of man (Fig. 13).

Pleasure, Joy, Affection.—Young chimpanzees make a kind of barking noise when pleased. When this noise, which the keepers call a laugh, is uttered, the lips are protruded. The form of the lips when they were pleased differed a little from that assumed when they were angered. If a young chimpanzee be tickled —and the armpits are particularly sensitive to tickling, as in the case of our children,—a more decided chuckling or laughing sound is uttered. The corners of the mouth are then drawn backwards; and this sometimes causes the lower eyelids to be slightly wrinkled. This wrinkling, so characteristic of our own laughter, is even more plainly seen in the monkeys (Fig. 13). The teeth in the upper jaw in the chimpanzee are not exposed when they utter their laughing noise, in which respect they differ from us, but their eyes sparkle and grow brighter.

Young Orangs, when tickled, likewise grin and make a chuckling sound and the eyes grow brighter. As soon as their laughter ceases an expression passes over their faces which, as Mr. Wallace remarked to me, may be called a smile. I have also noticed something of the same kind with the chimpanzee. Dr. Duchenne kept a monkey in his house, and when he gave it some choice delicacy an expression of satisfaction, partaking of the nature of an incipient smile, resembling that of man, could be plainly perceived.

[*Daily Mail.*

FIG. 13.—The remarkably human-like expression in this young Chimpanzee bears out the general principle that the structural differences between the young of man and ape are less than between the full-grown of these species. Such differences get less and less in pre-birth stages until, in the earliest stages, the embryos are indistinguishable.

A *Cebus* monkey, when rejoiced at again seeing a beloved person, utters a peculiar tittering sound. It also expresses agreeable sensation by drawing back the corners of its mouth without producing any sound, a movement that might appropriately be called a smile. The form of the mouth is different when pain or terror is expressed by high shrieks. *Cebus hypoleucus*, when pleased, makes a reiterated shrill note, and likewise draws back the corners of its mouth. So does the Barbary ape to an extraordinary degree, with, in addition, considerable wrinkling of the lower eyelids. At the same time it rapidly moves its lips in a spasmodic manner, the teeth being exposed; the noise produced being hardly more distinct than that called silent laughter. Two of the keepers in the Zoological Gardens affirmed that this slight sound was the animal's laughter, and when I expressed some doubt on this head they made it threaten a hated Entellus monkey living in the same compartment. Instantly the whole expression of the face of the Barbary ape changed; the mouth was opened more widely, the canine teeth more fully exposed, and a barking noise was uttered.

An Anubis baboon was put into a furious rage by his keeper, who then made friends with him and shook hands. As the reconciliation was effected the baboon rapidly moved his jaws and lips up and down, and looked pleased. When we laugh heartily a similar movement may be observed more or less distinctly in our jaws; though it is principally the muscles of the chest that are acted on, whilst with this baboon, and with some other monkeys, it is the muscles of the jaws and lips which are spasmodically affected.

With *Cynopithecus*, the corners of the mouth are at the same time drawn backwards and upwards, so that the teeth are exposed. Hence this expression would never be recognized by a stranger as one of pleasure. The crest of long hairs on the forehead is depressed, and apparently the whole skin of the head drawn backwards. The eyebrows are thus raised a little, and the eyes assume a staring appearance, while the lower eyelids become slightly wrinkled.

Painful Emotions and Sensations.—With monkeys the expression of slight pain, or of painful emotion, such as grief, vexation, jealousy, etc., is not easily distinguished from that of moderate anger; and these states of mind readily and quickly pass into each other. Grief, however, with some species is certainly exhibited by weeping. A woman, who sold a monkey, believed to have come from Borneo, to the Zoological Society, said that it often cried; and the keeper has repeatedly seen it weeping so copiously that the tears rolled down its cheeks. Two other specimens, believed to be the same species, have never been seen to weep, though they were carefully observed by the keeper and myself when much distressed and loudly screaming. Rengger states that the eyes of the *Cebus azaræ* fill with tears, but not sufficiently to overflow, when it is prevented getting some much-desired object, or is much frightened. Humboldt also asserts that the eyes of *Callithrix sciureus* "instantly fill with tears when it is seized with fear"; but when this pretty little monkey in the Zoological Gardens was teased, so as to cry out loudly, this did not occur.

The appearance of dejection in young orangs and chimpanzees when out of health, is as plain and almost as pathetic as in the case of our children. This state of mind and body is shown by their listless movements, fallen countenances, dull eyes, and changed complexion.

Anger.—"Some species, when irritated, pout the lips, gaze with a fixed and savage glare on their foe, and make repeated short starts as if about to spring forward, uttering at the same time inward guttural sounds. Many display their anger by suddenly advancing, making abrupt starts, at the same time opening the mouth and pursing up the lips so as to conceal the teeth, while the eyes are daringly fixed on the enemy as if in savage defiance. Some again, and principally the long-tailed monkeys or Guenons, display their teeth, and accompany their malicious grins with a sharp, abrupt, reiterated cry."[1] Some species uncover their teeth when enraged, whilst others conceal them by the protrusion of their lips; and some kinds draw back their ears. *Cynopithecus niger* acts in this manner, at the same time depressing the crest of hair on its forehead and showing its teeth; so that the movements of the features from anger are nearly the same as those from pleasure.

Two baboons, when first placed in the same compartment, sat opposite to each other and alternately opened their mouths; and this action seemed to end in a real yawn. Both animals seemed to wish to show that they are provided with a formidable set of teeth, as was undoubtedly the case. Some species of Macacus and of Cercopithecus behave in the same manner. Baboons seem to act consciously when they threaten by opening their mouths, for specimens with their canine teeth sawn off never acted in this manner; they would not show their comrades that they were powerless (Fig. 12). Baboons likewise show their anger by striking the ground with one hand, just as an angry man strikes the table with his fist.

The face of *Macacus rhesus*, when much enraged, grows red like that of a man in a violent passion. When the Mandrill is annoyed the brilliantly coloured, naked parts of the skin become still more vividly coloured.

With several species of baboons the ridge of the forehead projects much over the eyes and is studded with a few long hairs, representing the eyebrows. These animals are always looking about them, and in order to look upwards they have acquired the habit of frequently moving their eyebrows. Now many kinds of monkeys, especially baboons, when

[1] W. L. Martin.

angered or in any way excited, rapidly and incessantly move their eyebrows up and down, as well as the hairy skin of their foreheads.

The lips of young orangs and chimpanzees are protruded, sometimes to a wonderful degree, under

FIG. 14.—Chimpanzee disappointed and sulky because an orange had been taken away from him. A similar, though slighter, pouting may be seen in sulky children.

various circumstances. They act thus, not only when slightly angered, sulky, or disappointed, but when alarmed, and likewise when pleased. But neither the degree of protrusion nor the shape of the mouth is exactly the same in all cases; and the sounds which are then uttered are different. Fig. 14 represents a chimpanzee made sulky by an orange having been offered him, and then taken away. A similar protrusion or pouting of the lips, though to a much slighter degree, may be seen in sulky children.

When we try to perform some little action which is difficult and requires precision, for instance, to thread a needle, we generally close our lips firmly in order not to disturb our movements by breathing; and I noticed the same action in a young orang. The poor little creature was sick, and was amusing itself by trying to kill the flies on the window-panes with its knuckles; this was difficult as the flies buzzed about, and at each attempt the lips were firmly compressed and at the same time slightly protruded.

Although the countenances and gestures of orangs and chimpanzees are in some respects highly expressive (Fig. 13), I doubt whether on the whole they are so expressive as those of some other kinds of monkeys. This may be attributed in part to their ears being immovable, and in part to the nakedness of their eyebrows, of which the movements are thus rendered less conspicuous. When, however, they raise their eyebrows their foreheads become, as with us, transversely wrinkled. In comparison with man their faces are inexpressive, chiefly owing to their not frowning under emotion. Frowning, which is one of the most important of all the expressions in man, is due to the contraction of the muscles by which the eyebrows are lowered and brought together, so that vertical furrows are formed on the forehead (Fig. 22). Both the orang and chimpanzee[1] possess these muscles, but they seem rarely brought into action, at least in a conspicuous manner. I made my hands into a sort of cage, and placing some tempting fruit within, allowed both a young orang and chimpanzee to try their utmost to get it out; but although they grew rather cross, they showed not a trace of a frown. Nor was there any frown when they were enraged. Twice I took two chimpanzees from their rather

[1] Professor Macalister states that in the chimpanzee the *corrugator supercilii* is inseparable from the *orbicularis palpebrarum.*

dark room suddenly into bright sunshine, which would certainly have caused us to frown; they blinked and winked their eyes, but only once did I see a very slight frown. On another occasion I tickled the nose of a chimpanzee with a straw, and as it crumpled up its face slight vertical furrows appeared between the eyebrows. I have never seen a frown on the forehead of the orang.

The gorilla, when enraged, is described as erecting its crest of hair, throwing down its under lip, dilating its nostrils, and uttering terrific yells. The great power of movement in the scalp of the gorilla, of many baboons and other monkeys, deserves notice in relation to the power possessed by some few men, either through reversion or persistence, of voluntarily moving their scalps.[1]

Astonishment, Terror.—A live turtle was placed in the same compartment with many monkeys; and they showed unbounded astonishment, as well as some fear. They remained motionless, staring intently, their eyebrows being often moved up and down. Their faces seemed somewhat lengthened. They occasionally raised themselves on their hind-legs to get a better view. They often retreated a few feet, and then, turning their heads over one shoulder, again stared intently. It was curious to observe how much less afraid they were of the turtle than of a living snake which I had formerly placed in their compartment, for in the course of a few minutes some of the monkeys ventured to approach and touch the turtle. On the other hand, some of the larger baboons were greatly terrified, and grinned as if on the point of screaming out. When I showed a little dressed-up doll to *Cynopithecus niger*, it stood motionless, stared intently with widely opened eyes, and advanced its ears a little forwards. But when the turtle was placed in its compartment, this monkey also moved its lips in an odd, rapid, jabbering manner.

Attention, which precedes astonishment, is ex-

[1] See *Descent of Man.*

pressed by man by a slight raising of the eyebrows; and Dr. Duchenne informs me that when he gave to a monkey some new article of food, it elevated its eyebrows a little, thus assuming an appearance of close attention. It then took the food in its fingers, and, with lowered eyebrows, scratched, smelt, and examined it—an expression of reflection being thus exhibited. Sometimes it would throw back its head a little, and again with suddenly raised eyebrows re-examine and finally taste the food.

In no case did any monkey keep its mouth open when it was astonished. Mr. Sutton observed for me a young orang and chimpanzee during a considerable length of time; and however much they were astonished, or whilst listening intently to some strange sound, they did not keep their mouths open. This fact is surprising, as with mankind hardly any expression is more general than a widely open mouth under the sense of astonishment. Monkeys breathe more freely through their nostrils than men do; and this may account for their not opening their mouths when they are astonished; man apparently acts in this manner when startled, at first for the sake of quickly drawing a full inspiration, and afterwards for the sake of breathing as quietly as possible.

Terror is expressed by many kinds of monkeys by the utterance of shrill screams; the lips being drawn back, so that the teeth are exposed. Their hair becomes erect, especially when some anger is likewise felt, the face may turn pale, there is trembling, the excretions may be voided, and I have seen one which almost fainted from an excess of terror.

CHAPTER VI

SPECIAL EXPRESSIONS OF MAN: SUFFERING AND WEEPING

Weeping.—Signs of extreme pain, as shown by screams or groans, with the writhing of the whole body and the teeth clenched or ground together, are often accompanied or followed by profuse sweating, pallor, trembling, utter prostration, or faintness.

Infants, when suffering even slight pain, moderate hunger, or discomfort, utter violent and prolonged screams. Whilst thus screaming their eyes are firmly closed, so that the skin round them is wrinkled and the forehead contracted into a frown. The mouth is widely opened, with the lips retracted in a peculiar manner, which causes it to assume a squarish form, the gums or teeth being more or less exposed. The breath is inhaled spasmodically.

The firm closing of the eyelids and consequent compression of the eyeball serves to protect the eyes from becoming too much gorged with blood. The corrugators (*c. supercilii*) seem to be the first muscles to contract; and these draw the eyebrows downwards and inwards towards the base of the nose, causing a frown to appear between the eyebrows; at the same time they cause the disappearance of the transverse wrinkles across the forehead. The orbicular muscles contract almost simultaneously with the corrugators, and produce wrinkles all round the eyes. Lastly, the pyramidal muscles of the nose contract; and these draw the eyebrows and the skin of the forehead still lower down, producing short transverse wrinkles across the base of the nose. For the sake of brevity these muscles will generally be spoken of as the orbiculars, or as those surrounding the eyes. When they are strongly contracted, those running to the upper lip [1] likewise contract and raise it.

The raising of the upper lip draws upwards the flesh of the upper parts of the cheeks, and produces on each cheek a strongly-marked fold—the naso-labial—which runs from near the wings of the nostrils to the corners of the mouth and below them. This fold or furrow is very characteristic of the expression of a crying child; though a nearly similar fold is produced in the act of laughing or smiling.

As the upper lip is much drawn up during the act of screaming, the depressor muscles of the angles of the mouth are strongly contracted in order to keep the mouth widely open so that a full volume of sound may be poured forth. The action of these opposed muscles, above and below, tends to give to the mouth an oblong, almost squarish outline. An excellent observer,[2] in describing a baby crying whilst being fed, says, "It made its mouth like a square, and let the porridge run out at all four corners."

Infants whilst young do not shed tears or weep, as is well known to nurses and medical men. This circumstance is not exclusively due to the lacrymal glands being as yet incapable of secreting tears. I first noticed this fact from having accidentally brushed with the cuff of my coat the open eye of one of my infants, when seventy-seven days old, causing this eye to water freely; and though the child screamed violently, the other eye remained dry, or was only slightly suffused with tears. A similar slight effusion occurred ten days previously in both eyes during a screaming-fit. It would appear as if the lacrymal glands required some practice in the individual before they are easily excited into action, in somewhat the same manner as various inherited consensual movements and tastes require some exercise before they are fixed and perfected. This is all the more likely

[1] The *levator labii superioris alæque nasi*, the *levator labii proprius*, the *malaris*, and the *zygomaticus minor*.

[2] Mrs. Gaskell, *Mary Barton*, new edit. p. 84.

with a habit like weeping, which must have been acquired since the period when man branched off from the common progenitor of the genus *Homo* and of the non-weeping anthropomorphous apes.

A lady informs me that her child, nine months old, when in a passion screams loudly but does not weep; tears, however, are shed when she is punished by her chair being turned with its back to the table. This difference may perhaps be attributed to weeping being restrained at a more advanced age under most circumstances excepting grief.

With adults, especially of the male sex, weeping soon ceases to be caused by, or to express, bodily pain. This may be accounted for by its being thought weak and unmanly by men, both of civilized and barbarous races, to exhibit bodily pain by any outward sign. With this exception, savages weep copiously from very slight causes. A New Zealand chief cried like a child because some sailors spoilt his favourite cloak. I saw in Tierra del Fuego a native who had lately lost a brother, and who alternately cried with hysterical violence and laughed heartily at anything which amused him. Englishmen rarely cry, except under the pressure of the acutest grief; whereas in some parts of the Continent the men shed tears readily and freely.

The insane notoriously give way to all their emotions, and nothing is more characteristic of melancholia, even in the male sex, than a tendency to weep on the slightest occasions. Persons born idiotic likewise weep; but this is not the case with cretins, who never shed tears, but merely howl and shriek on occasions which would naturally produce weeping.

Weeping seems to be the primary and natural expression, as we see in children, of suffering of any kind, whether bodily pain short of extreme agony, or mental distress. But the foregoing facts and common experience show us that a frequently repeated effort to restrain weeping, in association with certain states of the mind, does much in checking the habit. On the other hand, it appears that the power of weeping can be increased through habit. The native women of New Zealand can voluntarily shed tears in abundance; they meet for this purpose to mourn for the dead, and they take pride in being able to cry in a most affecting manner.

Sobbing seems to be peculiar to the human species; keepers in the Zoological Gardens have never heard a real sob from any of the monkeys. We thus see that there is a close analogy between sobbing and the free shedding of tears; for with children sobbing does not commence during early infancy, but afterwards comes on rather suddenly, and then follows every bad crying-fit until the habit is checked with advancing years.

Contraction of Muscles round the Eyes during Screaming.—Infants and young children, whilst screaming, invariably close their eyes firmly by the contraction of the surrounding muscles, so that the skin becomes wrinkled all around. With older children, and even with adults, whenever there is violent and unrestrained crying, a tendency to the contraction of these same muscles may be observed; though this is often checked in order not to interfere with vision.

Not only are the muscles round the eyes strongly contracted during screaming, loud laughter, coughing, and sneezing, but during several other analogous actions. A man contracts these muscles when he violently blows his nose. I asked one of my boys to shout as loudly as he could, and as soon as he began, he firmly contracted his orbicular muscles; on asking him why he had so firmly closed his eyes I found that he was quite unaware of the fact—he had acted instinctively.

Chaucer, in describing a cock crowing, wrote:

> "This chaunteclere stood high upon his toos,
> Stretching his necke and held his eyen cloos,
> And gan to crowen loude for the nones."

It is not necessary in order to contract these

muscles that air should actually be expelled from the chest; it suffices that the muscles of the chest and abdomen should contract with great force, whilst by the closure of the glottis no air escapes. In violent vomiting or retching the diaphragm is made to descend; it is then held in this position while the glottis is closed. The abdominal muscles, as well as those of the stomach, now contract strongly and the contents of the stomach are thus ejected. During vomiting the head becomes congested, the features red and swollen, and the large veins of the face and temples dilate. At the same time the muscles round the eyes are strongly contracted.

The greatest exertion of the muscles of the body, if those of the chest are not brought into strong action in expelling or compressing the air within the lungs, does not lead to the contraction of the muscles round the eyes. I have observed my sons using great force in gymnastic exercises, but there was hardly any trace of contraction in the muscles round the eyes.

A high authority on the structure of the eye has shown that during violent expiration the external, the intra-ocular, and the retro-ocular vessels of the eye are all affected in two ways—namely, by the increased pressure of the blood in the arteries, and by the return of the blood in the veins being impeded. Both the arteries and the veins of the eye are more or less distended during violent expiration. We see the effects on the veins of the head, in their prominence, and in the purple colour of the face of a man who coughs violently. The whole eye advances a little during each violent expiration. This is due to the dilatation of the retro-ocular vessels. This also, I presume, is the reason that the eyes of a strangled man appear as if they were starting from their sockets.

With respect to the protection of the eye (during violent expiratory efforts) by the pressure of the eyelids, Professor Donders concludes that this action limits the dilatation of the vessels. At such times we not infrequently see the hand involuntarily laid upon the eyelids, as if the better to support and defend the eyeball. Professor Donders adds, "After injury to the eye, after operations, and in some forms of internal inflammation, we attach great value to the uniform support of the closed eyelids, and we increase this in many instances by the application of a bandage. In both cases we carefully endeavour to avoid great expiratory pressure, the disadvantage of which is well known."

Forcible expiratory efforts sometimes rupture the little external blood-vessels of the eye. With respect to the internal vessels, Dr. Gunning has recorded a case of exophthalmos [1] in consequence of whooping-cough, which depended on the rupture of the deeper vessels. We may, therefore, safely conclude that the firm closure of the eyelids during the screaming of children is an action full of meaning and of real service.

Dogs and cats in crunching bones, and sometimes in sneezing, close their eyelids; though dogs do not do so whilst barking loudly. Mr. Sutton found that a young orang and chimpanzee always closed their eyes in sneezing and coughing, but not whilst screaming. I gave a small pinch of snuff to a Cebus monkey and it closed its eyelids whilst sneezing; but not on a subsequent occasion whilst uttering loud cries.

Cause of the Secretion of Tears.—According to Henle there is a general tendency for the symptoms of emotional states to begin near the head and spread downwards. As an example he points out that in terror the sweat first breaks out on the forehead. In the same way he says that in strong emotion the flow of tears is the first effect, then follows the saliva, and in still more violent mental states the liver and other abdominal viscera are affected. Henle relies on anatomy entirely, for he says, "If by bad luck the origin of the nerve which excites the salivary

[1] Protrusion forward of the eyeball.—C. M. B.

gland had been nearer to the cerebral hemispheres than the nerves of lacrymation, poets must have celebrated salivation instead of weeping." This kind of generalization leaves unexplained the *specific* action of different emotions—why should we not perspire with grief, instead of with terror? Whenever the muscles round the eyes are strongly contracted to protect the eyes, tears are secreted, often in sufficient abundance to roll down the cheeks. This occurs under the most opposite emotions, and under no emotion at all. The sole exception, and this is only a partial one, to the existence of a relation between the contraction of these muscles and the secretion of tears, is that of young infants, who, whilst screaming violently with their eyelids firmly closed, do not commonly weep until they have attained the age of from two to three or four months. Their eyes, however, become suffused with tears at a much earlier age.

Under the opposite emotion of great joy or amusement, so long as laughter is moderate, there is hardly any contraction of the muscles round the eyes, so that there is no frowning; but when peals of loud laughter are uttered, with rapid and violent spasmodic expirations, tears stream down the face. I have more than once noticed that after a paroxysm of violent laughter the orbicular muscles and those running to the upper lip were still partially contracted, which together with the tear-stained cheeks gave to the upper half of the face an expression not to be distinguished from that of a child blubbering from grief. Tears streaming down the face during violent laughter is common to all the races of mankind.

Yawning commences with a deep inspiration, followed by a long and forcible expiration; and at the same time almost all the muscles of the body are strongly contracted, including those round the eyes. During this act tears are often secreted, and I have seen them even rolling down the cheeks.

When persons scratch some point which itches

intolerably, they forcibly close their eyelids, but I have never noticed that the eyes then become filled with tears. Such forcible closure of the eyelids is quite different from the gentle closure of the eyes which often accompanies the smelling a delicious odour, or the tasting a delicious morsel, and which probably originates in the desire to shut out any disturbing impression through the eyes.

The Indian elephant is said sometimes to weep. Sir E. Tennent, in describing those which he saw captured and bound in Ceylon, says, some "lay motionless on the ground, with no other indication of suffering than the tears which suffused their eyes and flowed incessantly." Speaking of another elephant, he says, "When overpowered and made fast, his grief was most affecting; his violence sank to utter prostration, and he lay on the ground, uttering choking cries, with tears trickling down his cheeks." In the Zoological Gardens the keeper of the Indian elephants positively asserts that he has several times seen tears rolling down the face of the old female, when distressed by the removal of the young one. Hence I was extremely anxious to ascertain, as an extension of the relation between the contraction of the orbicular muscles and the shedding of tears in man, whether elephants when screaming or trumpeting loudly contract these muscles. At Mr. Bartlett's desire the keeper ordered the old and the young elephant to trumpet; and we repeatedly saw in both animals that, just as the trumpeting began, the orbicular muscles, especially the lower ones, were distinctly contracted. On a subsequent occasion the keeper made the old elephant trumpet much more loudly, and invariably both the upper and lower orbicular muscles were strongly contracted, and now in an equal degree. The African elephant, which is so different from the Indian species that it is placed in a sub-genus, when made on two occasions to trumpet loudly, exhibited no trace of the contraction of the orbicular muscles.

The primary function of the secretion of tears is to cleanse and lubricate the surface of the eye; and a secondary one is to keep the nostrils damp, so that the inhaled air may be moist, and likewise to favour the power of smelling. Another, and equally important function of tears, is to wash out particles of dust or other minute objects which may get into the eyes. That this is of great importance is clear from the cases in which the cornea has been rendered opaque through inflammation caused by particles of dust not being removed. The secretion of tears from some irritant in the eye is a reflex action; the foreign body irritates a peripheral nerve which sends an impression to certain sensory nerve-cells; these transmit an influence to other cells, and these again to the lacrymal glands. The influence transmitted to these glands causes the relaxation of the muscular coats of the smaller arteries; thus more blood permeates the glandular tissue, and this induces free secretion of tears. When the small arteries of the face, including those of the retina, are relaxed during a blush, the lacrymal glands are sometimes affected in a like manner, for the eyes become suffused with tears.

As soon as some primordial form became semi-terrestrial in its habits it was liable to get particles of dust into its eyes; if these were not washed out they would cause much irritation, and on the principle of the radiation of nerve-force to adjoining nerve-cells, the lacrymal glands would be stimulated to secretion. As this would often recur, and as nerve-force readily passes along accustomed channels, a slight irritation would ultimately suffice to cause a free secretion of tears.

As soon as by this, or by some other means, a reflex action of this nature had been established and rendered easy, other stimulants applied to the surface of the eye—such as a cold wind, inflammatory action, or a blow on the eyelids—would cause a copious secretion of tears. The glands are also excited into action through the irritation of adjoining parts. Thus when the nostrils are irritated by pungent vapours, though the eyelids may be kept firmly closed, tears are copiously secreted; and this likewise follows from a blow on the nose. A stinging switch on the face produces the same effect. In these latter cases the secretion of tears is an incidental result, and of no direct service. As all these parts of the face, including the lacrymal glands, are supplied with branches of the same nerve, namely, the fifth, it is intelligible that the effects of the excitement of one branch should spread to the other branches.

A strong light acting on the retina, when in a normal condition, has very little tendency to cause lacrymation; but with small, old-standing ulcers on the cornea, the retina becomes excessively sensitive to light, and exposure even to common daylight causes forcible and sustained closure of the lids, and a profuse flow of tears. When persons who ought to begin the use of convex glasses habitually strain the waning power of accommodation, an undue secretion of tears often follows.

The eye and adjoining parts are subject to an extraordinary number of reflex and associated movements, besides those relating to the lacrymal glands. When a bright light strikes the retina of one eye alone the iris contracts, but the iris of the other eye also contracts. The iris likewise moves in accommodation to near or distant vision, and when the two eyes are made to converge. Every one knows how irresistibly the eyebrows are drawn down under an intensely bright light (Fig. 15). The eyelids also involuntarily wink when an object is moved near the eyes, or a sound is suddenly heard. The well-known case of a bright light causing some persons to sneeze is even more curious; for nerve-force here radiates from certain nerve-cells in connection with the retina to the sensory nerve-cells of the nose, causing it to tickle; and from these to cells which command

[Fox Photos.

FIG. 15.—Little inmates of a Children's Home at Leytonstone, Essex, lined up for their daily "dose" of sunlight cure. Note the "screwing-up" of the eyes to protect the sensitive retinas

orbiculars and the respiratory muscles which expel the air through the nostrils.

Why are the tears secreted during a screaming-fit or violent expiratory efforts? As a slight blow on the eyelids causes a copious secretion of tears, it is possible that the spasmodic contraction of the eyelids, by pressing strongly on the eyeball, should in a similar manner cause some secretion. This seems possible, although the voluntary contraction of the same muscles does not produce any such effect. We know that a man cannot voluntarily sneeze or cough with nearly the same force as he does automatically; and so it is with the contraction of the orbicular muscles: Sir C. Bell experimented on them, and found that by suddenly and forcibly closing the eyelids in the dark sparks of light are seen, like those caused by tapping the eyelids with the fingers; "but in sneezing the compression is both more rapid and more forcible, and the sparks are more brilliant." That these sparks are due to the contraction of the eyelids is clear, because if they "are held open during the act of sneezing, no sensation of light will be experienced." Some weeks after the eye has been slightly injured, spasmodic contractions of the eyelids ensue accompanied by a flow of tears. In the act of yawning, the tears are apparently due solely to the spasmodic contraction of the muscles round the eyes. Notwithstanding these latter cases, it seems hardly credible that such pressure of the eyelids on the surface of the eye should be sufficient to cause the secretion of tears during violent expiratory efforts. Another cause may come conjointly into play. The internal parts of the eye, under certain conditions, act in a reflex manner on the lacrymal glands. During violent expiratory efforts the pressure of blood within the vessels of the eye is increased and the return of the venous blood is impeded. It seems, therefore, not improbable that the distension of the ocular vessels, thus induced, might react on the lacrymal glands and so augment the flow of tears due to the spasmodic pressure of the eyelids on the surface of the eye.

We should bear in mind that the eyes of infants have been acted on in this double manner during numberless generations whenever they have screamed; and on the principle of nerve-force readily passing along accustomed channels, even moderate compression of the eyeballs and moderate distension of the ocular vessels would ultimately come, through habit, to act on the glands. We have an analogous case in the orbicular muscles being almost always contracted in some slight degree, even during a gentle crying-fit, when there can be no distension of the vessels and no uncomfortable sensation excited within the eyes.

Moreover, when complex actions or movements have long been performed in strict association together, and these are from any cause at first voluntarily and afterwards habitually checked, then, if the proper exciting conditions occur, any part of the action or movement which is least under voluntary control will often still be involuntarily performed. Secretion by a gland is remarkably free from the influence of the will; therefore, when with advancing age of the individual, or with advancing culture of the race, the habit of crying out or screaming is restrained, and there is consequently no distension of the blood-vessels of the eye, it may nevertheless well happen that tears should still be secreted. The muscles round the eyes of a person who reads a pathetic story may twitch or tremble in so slight a degree as hardly to be detected. There has been no screaming and no distension of the blood-vessels, yet through habit certain nerve-cells send a small amount of nerve-force to the cells commanding the muscles round the eyes and to those commanding the lacrymal glands, for the eyes often become at the same time moist with tears. Had the twitchings and tears been prevented, it is nevertheless almost certain that there would have been some tendency to transmit nerve-force in these same directions; and as the lacrymal glands are free from all control of the will, they would be eminently liable still to act, thus betraying the pathetic thoughts passing through the person's mind.

If, during early life, when habits of all kinds are readily established, our infants, when pleased, had been accustomed to utter loud peals of laughter (during which the vessels of their eyes would have been distended) as often and as continuously as they have yielded when distressed to screaming-fits, then it is probable that in after life tears would have been as copiously and as regularly secreted under the one state of mind as under the other. Gentle laughter, or a smile, or even a pleasing thought, would have sufficed to cause a moderate secretion of tears. So again if infants had for many generations frequently suffered from prolonged choking-fits, during which the vessels of the eye were distended and tears secreted, then it is probable, such is the force of associated habit, that during after life the mere thought of a choke would have sufficed to bring tears to the eyes.

To sum up. Children, when suffering in any way, cry out loudly like the young of most other animals, partly as a call to their parents and partly because the exertion serves as a relief. Prolonged screaming inevitably leads to the gorging of the blood-vessels of the eye; and this will have led to the contraction of the muscles round the eyes in order to protect them. At the same time the spasmodic pressure on the surface of the eye, and the distension of the vessels within it, will have reflexly affected the lacrymal glands. Finally, through the three principles—of nerve-force readily passing along accustomed channels, of association, and of certain actions being more controlled by the will than others—it has come to pass that suffering readily causes the secretion of tears, without being necessarily accompanied by any other action.

In accordance with this view we must look at weeping as an incidental result, as purposeless as the secretion of tears from a blow on the eye, or as a sneeze from a bright light. Yet this does not present any difficulty in our understanding how the secretion of tears relieves suffering. And by as much as the weeping is more violent or hysterical, by so much will the relief be greater, on the same principle that the writhing of the body, the grinding of the teeth, and shrieking, all give relief during agony.

Chapter VII

LOW SPIRITS, ANXIETY, GRIEF, DEJECTION, DESPAIR

After the mind has suffered from a paroxysm of grief and the cause still continues, we fall into a state of low spirits or are utterly dejected. Prolonged pain generally leads to the same state of mind. If we expect to suffer, we are anxious; if we have no hope of relief, we despair.

Persons suffering from excessive grief often seek relief by violent and almost frantic movements; but when their suffering is somewhat mitigated, yet prolonged, they remain motionless and passive, or occasionally rock themselves to and fro. The circulation becomes languid, the face pale, the muscles flaccid, the eyelids droop, the head hangs on the contracted chest, the lips, cheeks, and lower jaw all sink downwards. Hence the features are lengthened; and the face of a person who hears bad news is said to fall. Native Tierra del Fuegians endeavoured to explain to us that the captain of a sealing vessel was out of spirits by pulling down their cheeks with both hands, so as to make their faces as long as possible. After prolonged suffering the eyes become dull and lack expression, and are often slightly suffused with tears. The eyebrows not rarely are rendered oblique, due to their inner ends being raised. This produces peculiarly formed wrinkles on the forehead, very different from those of a simple frown; though in some cases a frown alone may be present. The corners of the mouth are drawn downwards (Fig. 16), which is so universally recognized as a sign of being out of spirits, that it is proverbial. The breathing and circulation become slow and feeble, and the former is often interrupted by characteristic deep sighs. As the grief of a person in this state occasionally recurs and increases into a paroxysm, spasms affect the respiratory muscles, and he feels as if something, the so-called *globus hystericus*, was rising in his throat;[1] such movements are clearly allied to the sobbing of children, and to those severer spasms when a person is said to choke from excessive grief.

Obliquity of the Eyebrows.—The eyebrows may occasionally be seen to assume an oblique position in persons suffering from deep dejection or anxiety; for instance, I have observed this movement in a mother whilst speaking about her sick son; and it is sometimes excited by quite trifling or momentary causes of real or pretended distress. The eyebrows assume this position owing to the contraction of the orbicular, corrugator, and pyramidal muscles of the nose, which together lower and contract the eyebrows, being partially checked by the more powerful action of the central portion of the frontal muscle. The contraction of the latter raises the inner ends alone of the eyebrows; and as the corrugators at the same time draw the eyebrows together, their inner ends become puckered into a fold. The eyebrows are at the same time somewhat roughened, owing to the hairs being made to project. Dr. J. Crichton Browne has also often noticed in melancholic patients who keep their eyebrows persistently oblique, "a peculiar acute arching of the upper eyelid." This arching of the eyelids depends, I believe, on the inner end alone of the eyebrows being raised; for when the whole eyebrow is elevated and arched, the upper eyelid follows in a slight degree the same movement.

But the most conspicuous result of the opposed contraction of the above-named muscles is exhibited

[1] The familiar "lump in the throat" or *globus hystericus* is in part due to contraction of the constrictors of the pharynx and in part to activation of the thyroid gland in the front of the neck.—C. M. B.

by the peculiar furrows formed on the forehead. These muscles, when thus in conjoint yet opposed action, may be called, for the sake of brevity, the grief-muscles. When a person elevates his eyebrows by the contraction of the whole frontal muscle, transverse wrinkles extend across the whole breadth of the forehead (Fig. 21); but in the present case the middle part alone is contracted; consequently transverse furrows are formed across the middle part alone of the forehead. The skin over the exterior parts of both eyebrows is at the same time drawn downwards and smoothed, by the contraction of the outer portions of the orbicular muscles. The eyebrows are likewise brought together through the simultaneous contraction of the corrugators; and this latter action generates vertical furrows (Fig. 22), separating the exterior and lowered part of the skin of the forehead from the central and raised part. The union of these vertical furrows with the central and transverse furrows produces a mark on the forehead resembling a horse-shoe or, more strictly, the three sides of a quadrangle. They are often conspicuous on the foreheads of adult persons, when their eyebrows are made oblique; but with young children, owing to their skin not easily wrinkling, mere traces of them can be detected.

Few persons can voluntarily act on their grief-muscles; but after repeated trials a considerable number succeed. The degree of obliquity in the eyebrows, whether assumed voluntarily or unconsciously, differs much in different persons. With some who apparently have unusually strong pyramidal muscles, the contraction of the central portion of the frontal muscle, although it may be energetic as shown by the quadrangular furrows on the forehead, does not raise the inner ends of the eyebrows, but only prevents their being lowered as much as otherwise they would have been. The grief-muscles are brought into action much more frequently by children and women than by men. They are rarely acted on, at least with grown-up persons, from bodily pain, but almost exclusively from mental distress.

Grief-muscles are not very frequently brought into play; and as the action is often momentary it easily escapes observation. Hence probably it is that this expression is very seldom alluded to in any work of fiction.

The ancient Greek sculptors were familiar with the expression, as shown in the statues of the Laocoon and Arrotino; but, as Duchenne remarks, they carried the transverse furrows across the whole breadth of the forehead, and thus committed a great anatomical mistake: this is likewise the case in some modern statues. It is, however, more probable that these wonderfully accurate observers intentionally sacrificed truth for the sake of beauty, than that they made a mistake; for rectangular furrows on the forehead would not have had a grand appearance in marble. The expression is not often represented in pictures by the old masters, but a lady who is perfectly familiar with it informs me that in Fra Angelico's "Descent from the Cross" it is clearly exhibited in one of the figures on the right-hand.

Dr. Crichton Browne informs me that the grief-muscles may constantly be seen in energetic action in cases of melancholia and that persistent furrows, due to their habitual contraction, are characteristic of the physiognomy of melancholics. A widow, aged fifty-one, fancied that she had lost all her viscera and that her whole body was empty. She wore an expression of great distress, and beat together her semi-closed hands rhythmically for hours. The grief-muscles were persistently contracted and the upper eyelids arched. This condition lasted for months; she then recovered and her countenance resumed its natural expression.

The expression of grief appears to be common to all races of mankind. The lady who told me of Fra Angelico's picture saw a negro towing a boat on the Nile, and as he encountered an obstruction, she observed his grief-muscles in strong action, with the middle of the forehead well wrinkled.

Mr. J. Scott, of the Botanic Gardens, Calcutta, observed during some time, himself unseen, a very young Dhangar woman from Nagpore nursing her baby who was at the point of death; and he distinctly saw the eyebrows raised at the inner corners, the eyelids drooping, the forehead wrinkled in the middle, the mouth slightly open, with the corners much depressed. He then came from behind a screen of plants and spoke to the poor woman, who started, burst into a bitter flood of tears, and besought him to cure her baby. A Hindustani man, who from illness and poverty was compelled to sell his favourite goat, after receiving payment repeatedly looked at the money in his hand and then at his goat, as if doubting whether he would not return it. He went to the goat, which was tied up ready to be led away, and the animal reared up and licked his hands. His eyes then wavered from side to side; his "mouth was partially closed, with the corners very decidedly depressed." At last the poor man seemed to make up his mind that he must part with his goat, and then, as Mr. Scott saw, the eyebrows became slightly oblique, with the characteristic puckering or swelling at the inner ends, but the wrinkles on the forehead were not present. The man stood thus for a minute, then, heaving a deep sigh, burst into tears, raised up his two hands, blessed the goat, turned round, and without looking again, went away.

Obliquity of the Eyebrows under Suffering.—Why should grief or anxiety cause the central part alone of the frontal muscle, together with the muscles round the eyes, to contract? Here we seem to have a complex movement for the sole purpose of expressing grief; and yet it is a comparatively rare expression. I believe the explanation is not so difficult as it at first appears. Dr. Duchenne gives a photograph of a young man who, when looking upwards at a strongly illuminated surface, involuntarily contracted his grief-muscles in an exaggerated manner. On a very bright day with the sun behind me, I met, whilst on horseback, a girl whose eyebrows, as she looked up at me, became extremely oblique, with the proper furrows on her forehead. I made three of my children, without giving them any clue to my object, look as long and as attentively as they could, at the summit of a tall tree standing against an extremely bright sky. With all three, the orbicular, corrugator, and pyramidal muscles were energetically contracted through reflex action, so that their eyes might be protected from the bright light (Fig. 15). But they tried their utmost to look upwards; and now a curious struggle, with spasmodic twitchings, could be observed between the central portion of the frontal muscle and the several muscles which serve to lower the eyebrows and close the eyelids. The involuntary contraction of the pyramidal caused the basal part of their noses to be transversely and deeply wrinkled. In one of the three children the whole eyebrows were momentarily raised and lowered by the alternate contraction of the whole frontal muscle and of the muscles surrounding the eyes, so that the whole breadth of the forehead was alternately wrinkled and smoothed. In the other two children the forehead became wrinkled in the middle part alone, rectangular furrows being thus produced; and the eyebrows were rendered oblique, with their inner extremities puckered and swollen. In both these cases the eyebrows and forehead were acted on under the influence of a strong light, in precisely the same manner as under the influence of grief or anxiety.

The pyramidal muscle serves to draw down the skin of the forehead between the eyebrows together with their inner extremities. The central portion of the frontal is the antagonist of the pyramidal; and if the action of the latter is to be specially checked, this central part must be contracted.[1]

[1] Electrical stimulation of these muscles confirms the

When children scream they contract the orbicular, corrugator, and pyramidal muscles, for the sake of compressing their eyes and thus protecting them from being gorged with blood. I therefore expected to find with children, that when they endeavoured either to prevent a crying-fit from coming on, or to stop crying, they would check the contraction of the above-named muscles, and consequently that the central part of the frontal muscle would be brought into play. But I soon found that the grief-muscles were very frequently brought into distinct action on these occasions. A little girl, a year and a half old, was teased by some other children, and before bursting into tears her eyebrows became decidedly oblique. With an older girl the same obliquity was observed, with the inner ends of the eyebrows plainly puckered; and at the same time the corners of the mouth were drawn downwards. As soon as she burst into tears the features all changed and this peculiar expression vanished. Again, after a little boy had been vaccinated, which made him scream and cry violently, the surgeon gave him an orange brought for the purpose, and this pleased the child much; as he stopped crying all the characteristic movements were observed, including the formation of rectangular wrinkles in the middle of the forehead. Lastly, I met on the road a little girl three or four years old, who had been frightened by a dog, and when I asked her what was the matter she stopped whimpering, and her eyebrows instantly became oblique to an extraordinary degree.

Here then we have the key to the problem why the central part of the frontal muscle and the muscles round the eyes contract in opposition to each other under the influence of grief. We have all of us, as infants, repeatedly contracted our orbicular, corrugator, and pyramidal muscles, in order to protect our

---

conclusion that the *pyramidalis nasi* is "the direct antagonist of the central portion of the occipito-frontal, and *vice versa*."

eyes whilst screaming; our progenitors before us have done the same during many generations; and though with advancing years we easily prevent, when feeling distressed, the utterance of screams, we cannot, from long habit, always prevent a slight contraction of the above-named muscles; nor indeed do we observe their contraction in ourselves, or attempt to stop it, if slight. But the pyramidal muscles seem to be less under the command of the will than the other related

FIG. 16.—Small boy "down in the mouth" after having been hurt by another boy. He burst out crying just after the photograph had been taken. Note the depression of the corners of the mouth.

muscles; and their contraction can be checked only by the antagonistic contraction of the central part of the frontal muscle. The result is the oblique drawing up of the eyebrows, the puckering of their inner ends, and the formation of rectangular furrows on the middle of the forehead. As children and women cry much more freely than men, and as grown-up persons of both sexes rarely weep except from mental distress, we can understand why the grief-muscles are more frequently seen in action with children and women than with men; and with adults of both sexes from mental distress alone. In all cases of distress our brains tend through long habit to send an order to certain muscles to contract, as if we were still infants on the point of screaming out; but this order we, through habit, are able partially to counteract.

On the Depression of the Corners of the Mouth.—This action is effected by the *depressores anguli oris*, the fibres of which diverge downwards, with the upper convergent ends attached round the angles of the

FIG. 17.—Same boy later. The expression may now well be taken for one of "slyness."

mouth, and to the lower lip a little way within the angles. The contraction of this muscle draws downwards and outwards the corners of the mouth, including the outer part of the upper lip, and even in a slight degree the wings of the nostrils (Figs. 16, 17). When the mouth is closed and this muscle acts, the commissure or line of junction of the two lips forms a curved line with the concavity downwards, and the lips themselves are generally somewhat protruded, especially the lower one.

The expression of low spirits, grief or dejection, due to the contraction of this muscle has been noticed by every one who has written on the subject. To say that a person is "down in the mouth," is synonymous with saying that he is out of spirits. The depression of the corners may often be seen with the melancholic insane. It has been observed with men belonging to various races, namely with Hindoos, the dark hill-tribes of India, Malays, and with the aborigines of Australia.

Dr. Duchenne concludes from his observations that the depressors of the corners of the mouth are some of the facial muscles least under the control of the will. This fact may indeed be inferred from the fact that infants when doubtfully beginning to cry, or endeavouring to stop crying, generally command all the other facial muscles more effectually than they do these depressors. Now as the depressors have been repeatedly brought into strong action during infancy in many generations, nerve-force will tend to flow to these muscles as well as to various other facial muscles whenever in after life a feeling of distress is experienced. But as the depressors are less under the control of the will than most of the other muscles they would often slightly contract whilst the others remained passive. It is remarkable how small a depression of the corners of the mouth gives to the countenance an expression of low spirits or dejection, an extremely slight contraction of these muscles being sufficient to betray this state of mind. An old lady with a comfortable but absorbed expression sat nearly opposite to me in a railway carriage. Whilst I was looking at her, I saw that her *depressores anguli oris* became slightly contracted; but as her countenance remained placid I reflected how easily one might be deceived. The thought had hardly occurred to me when I saw that her eyes suddenly became suffused with tears almost to overflowing, and her whole countenance fell. There could now be no doubt that some painful recollection, perhaps that of a long-lost child, was passing through her mind. As soon as her sensorium was thus affected, certain nerve-cells from long habit instantly transmitted an order to all the respiratory muscles, and to those round the mouth, to prepare for a fit of crying. But the order was countermanded by a later acquired habit, and all the muscles were obedient, excepting in a slight degree the muscles which draw down the corners of the mouth.

Through steps such as these we can understand why as soon as some melancholy thought passes through the brain, there occurs a just perceptible drawing down of the corners of the mouth and raising up of the inner ends of the eyebrows, and immediately afterwards a slight suffusion of tears. Such actions are vestiges of the screaming-fits of infancy. The links are indeed wonderful which connect cause and effect in giving rise to various expressions on the human countenance; and they explain to us the meaning of certain movements, which we involuntarily and unconsciously perform, whenever certain transitory emotions pass through our minds.

CHAPTER VIII

JOY, HIGH SPIRITS, LOVE, TENDER FEELINGS, DEVOTION

JOY, when intense, leads to various purposeless movements—to dancing about, clapping the hands, stamping, loud laughter, etc. Children at play are almost incessantly laughing. With young persons past childhood, when they are in high spirits, there is always much meaningless laughter. The laughter of the gods is described by Homer as "the exuberance of their celestial joy after their daily banquet." A man smiles—and smiling, as we shall see, graduates into laughter—at meeting an old friend in the street, as he does at any trifling pleasure, such as smelling a sweet perfume. Laura Bridgman, from her blindness and deafness, could not have acquired any expression through imitation, yet when a letter from a beloved friend was communicated to her by gesture-language, she "laughed and clapped her hands, and the colour mounted to her cheeks." On other occasions she has been seen to stamp for joy.

Certain idiots afford good evidence that laughter or smiling expresses mere happiness or joy. Among them laughter is the most prevalent and frequent of all the emotional expressions. Their countenances often exhibit a stereotyped smile; their joyousness is increased, and they grin, chuckle, or giggle, whenever food is placed before them, or when they are caressed, shown bright colours, or hear music. Most of these idiots are destitute of any distinct ideas: they simply feel pleasure and express it by laughter or smiles.

With normal grown-up persons laughter is excited by causes other than those which suffice during childhood; but this remark hardly applies to smiling. Laughter in this respect is analogous to weeping, which with adults is almost confined to mental distress, whilst with children it is excited by bodily suffering, as well as by fear or rage. Something incongruous, exciting surprise and some sense of superiority in the laugher, who must be in a happy frame of mind, seems to be the commonest cause among grown-up persons. The circumstances must not be of a momentous nature: no poor man would laugh or smile on suddenly hearing that a large fortune had been bequeathed to him. If the mind is strongly excited by pleasurable feelings, and any little unexpected event or thought occurs, a large amount of nervous energy discharges itself through the motor nerves to various muscles producing the half-convulsive actions we term laughter. During the siege of Paris the German soldiers, after strong excitement from exposure to extreme danger, were particularly apt to burst out into loud laughter at the smallest joke. Mr. C. Hinton, of San Francisco, describes himself as alternately screaming for help and laughing when alone in a position of great danger on the cliffs near the Golden Gate. So again when young children are just beginning to cry, an unexpected event will sometimes suddenly turn their crying into laughter, which apparently serves equally well to expend their superfluous nervous energy.

The imagination is sometimes said to be "tickled" by a ludicrous idea, and this so-called "tickling of the mind" is curiously analogous to that of the body. Every one knows how immoderately children laugh, and how their whole bodies are convulsed when they are tickled. The anthropoid apes likewise utter a reiterated sound, corresponding with our laughter, when they are tickled. I touched with a bit of paper the sole of the foot of one of my infants when only seven days old, and it was suddenly jerked away and the toes curled about as in an older child. Such movements, as well as laughter from being tickled, are manifestly reflex actions; and this is likewise shown by the minute unstriped muscles, which serve to erect the separate hairs on the body, contracting near a tickled surface. Yet laughter from a ludicrous idea, though involuntary, cannot strictly be called a reflex action. In this case, and in that of laughter from being tickled, the mind must be in a pleasurable condition; a young child, if tickled by a strange man, would scream from fear. The touch must be light, and an idea or event, to be ludicrous, must not be of grave import. The parts of the body which are most easily tickled are those which are not commonly touched, such as the armpits or between the toes, or the soles of the feet which are habitually touched by a broad surface. From the fact that a child cannot tickle itself in the same degree that another person can tickle it, it seems that the point to be touched must not be known; so with the mind, something unexpected—a novel or incongruous idea which breaks through an habitual train of thought—appears to be a strong element in the ludicrous. L. Dumont seeks to show that tickling depends on *unexpected* variations in the nature of the contact; he, too, believes that it is this unexpectedness which allies tickling with the ludicrous as a cause of laughter.

The sound of laughter is produced by a deep inspiration followed by short, interrupted, spasmodic contractions of the chest, and especially of the diaphragm. Hence we hear of "laughter holding both his sides." From the shaking of the body, the head nods to and fro. The lower jaw often quivers up and down, as is likewise the case with some species of baboons when they are much pleased.

During laughter the mouth is opened more or less widely, with the corners drawn much backwards, as well as a little upwards; and the upper lip is somewhat raised. The drawing back of the corners is best seen in moderate laughter, and especially in a broad smile—the latter epithet showing how the mouth is widened.

By the drawing backwards and upwards of the corners of the mouth, through the contraction of the great zygomatic muscles, and by the raising of the upper lip, the cheeks are drawn upwards. Wrinkles are thus formed under the eyes, and, with old people, at their outer ends; and these are highly characteristic of laughter or smiling. As a gentle smile increases into a laugh, the upper lip is drawn up and the lower orbiculars contract, the wrinkles beneath the eyes are strengthened or increased and at the same time the eyebrows are slightly lowered.

The contraction of the zygomatic muscles under pleasurable emotions is shown by patients suffering from *general paralysis of the insane.* "In this malady there is almost invariably optimism—delusions as to wealth, rank, grandeur—insane joyousness, benevolence, and profusion, while its very earliest physical symptom is trembling at the corners of the mouth and at the outer corners of the eyes. . . . The countenance has a pleased and benevolent expression."[1]

As in laughing and broadly smiling the cheeks and upper lip are much raised, the nose appears to be shortened, and the skin on the bridge becomes finely wrinkled in transverse lines, with oblique longitudinal lines on the sides. The upper front teeth are commonly exposed. A well-marked naso-labial fold (often double in old persons) is formed, which runs from the wing of each nostril to the corner of the mouth.

A bright and sparkling eye is as characteristic of a pleased or amused state of mind, as are the wrinkles produced by the retraction of the corners of the mouth and upper lip. Even the eyes of microcephalous idiots, who never speak, brighten slightly when they are pleased. Under extreme laughter the eyes are too suffused with tears to sparkle; but the moisture during moderate laughter or smiling may aid in giving them lustre, though this must be of sub-

[1] Dr. J. Crichton Browne.

ordinate importance as they become dull from grief, though then often moist. Their brightness seems to be chiefly due to their tenseness, owing to the contraction of the orbicular muscles and to the pressure of the raised cheeks. I remember seeing a man utterly prostrated by prolonged and severe exertion during a very hot day, and a bystander compared his eyes to those of a boiled codfish.

We can see in a vague manner how the utterance of sounds during laughter would naturally become associated with a pleasurable state of mind; for throughout a large part of the animal kingdom vocal or instrumental sounds are employed either as a call or as a charm by one sex for the other. They are also employed as the means for a joyful meeting between the parents and their offspring, and between the attached members of the same social community. They would naturally be as different as possible from the screams or cries of distress; and as in the production of the latter the expirations are prolonged and continuous, with the inspirations short and interrupted, so we should expect with the sounds of joy, that the expirations would have been short and broken, with the inspirations prolonged; and this is the case.

During a paroxysm of laughter the mouth is opened to its utmost extent, yet hardly any sound is emitted, or it seems to come from deep down in the throat. The respiratory muscles, and even those of the limbs, are at the same time thrown into rapid vibratory movements. We may infer that all these effects are due to some common cause, for all are characteristic and expressive of a pleased state of mind in monkeys as well as in ourselves.

A graduated series can be followed from violent to moderate laughter, to a broad smile, to a gentle smile, and to the expression of mere cheerfulness. During excessive laughter the whole body is often thrown backward and shakes, or is almost convulsed; the respiration is much disturbed; the head and face become gorged with blood, and the veins distended; and the orbicular muscles are spasmodically contracted in order to protect the eyes. Tears are freely shed, and hence it is scarcely possible to point out any difference between the tear-stained face of a person after a paroxysm of excessive laughter and after a bitter crying-fit. "It is curious to observe, and it is certainly true," Sir J. Reynolds remarks, "that the extremes of contrary passions are, with very little variation, expressed by the same action." He gives as an instance the frantic joy of a Bacchante and the grief of a Mary Magdalen. It is probably due to the close similarity of the spasmodic movements caused by these widely different emotions that hysteric patients alternately cry and laugh, and that children sometimes pass suddenly from the one to the other state.

Native women in the Malay peninsula sometimes shed tears when they laugh heartily. A common expression of the Dyaks of Borneo is: "we nearly made tears from laughter." The aborigines of Australia are described as jumping about and clapping their hands for joy, and as often roaring with laughter, their eyes freely watering on such occasions. They have a keen sense of the ridiculous and are excellent mimics. With Europeans hardly anything excites laughter so easily as mimicry, and it is significant to find the same fact with so distinct a race as the savages of Australia.

The eyes of South African Kaffirs, especially the women, often fill with tears during laughter. The painted face of a Hottentot woman has been seen furrowed with tears after a fit of laughter. With the Abyssinians tears are secreted under the same circumstances, and in North America the same fact has been observed in a remarkably savage and isolated tribe. Mr. B. F. Hartshorne states in the most positive manner that the Weddas of Ceylon never laugh. Every conceivable incentive to laughter was used in vain. When asked whether they ever laughed, they replied, "No, what is there to laugh at?" No abrupt line of demarcation can be drawn between the movement of the features during the most violent laughter and a very faint smile.

High Spirits, Cheerfulness.—A man in high spirits, though he may not actually smile, commonly exhibits some retraction of the corners of his mouth. From the excitement of pleasure the circulation becomes more rapid; the eyes are bright, and the colour of the face rises. The brain, stimulated by the increased flow of blood, reacts on the mental powers; lively ideas pass more rapidly through the mind, and the affections are warmed. I heard a child, a little under four years old, when asked what was meant by being in good spirits, answer, "Laughing, talking, and kissing." It would be difficult to give a truer and more practical definition. A man in this state holds his body erect, his head upright, and his eyes open. There is no drooping of the features, and no contraction of the eyebrows. The frontal muscle contracts slightly and smooths the brow, removes every trace of a frown, arches the eyebrows a little, and raises the eyelids. Hence the Latin phrase, *exporrigere frontem*—to unwrinkle the brow—means, to be cheerful or merry. The whole expression of a man in good spirits is the opposite of that of one suffering from sorrow. In exhilarating emotions the eyebrows, eyelids, the nostrils, and the angles of the mouth are raised; in depressing emotions it is the reverse. In joy the face expands, in grief it lengthens.

Savages sometimes express their satisfaction not only by smiling, but by gestures derived from the pleasure of eating. Petherick describes how the negroes on the Upper Nile began to rub their bellies when he displayed his beads; and Leichhardt says that the Australian natives smacked and clacked their mouths at the sight of his horses and bullocks, and more especially of his kangaroo dogs. When Greenlanders affirm anything with pleasure they suck down air with a certain sound; possibly in imitation of the act of swallowing savoury food.

Laughter is suppressed by the firm contraction of the orbicular muscles of the mouth, which prevents the great zygomatic and other muscles from drawing the lips backwards and upwards. The lower lip is also sometimes held by the teeth, and this gives a roguish expression to the face.

Laughter is frequently forced to mask some other state of mind, even anger. Persons often laugh to conceal shame or shyness. In derision, a real or pretended smile or laugh is often blended with the expression proper to contempt. Here the meaning of the laugh or smile is to show the offending person that he excites only amusement.

Love, Tender Feelings, etc.—Although love—for instance, that of a mother for her infant—is one of the strongest of the emotions, it can hardly be said to have any peculiar means of expression; and this is intelligible, as it has not habitually led to any special line of action. No doubt, as affection is a pleasurable sensation, it generally causes a gentle smile and some brightening of the eyes. A strong desire to touch the beloved person is commonly felt; and love is expressed by this means more plainly than by any other. "Tenderness is a pleasurable emotion variously stimulated," remarks Mr. Bain, "whose effort is to draw human beings into mutual embrace. Hence we long to clasp in our arms those whom we tenderly love. We probably owe this desire to inherited habit, in association with the nursing and tending of our children, and with the mutual caresses of lovers."

With the lower animals we see the same principle of pleasure derived from contact in association with love. Dogs and cats manifestly take pleasure in rubbing against their masters and mistresses, and in being rubbed or patted by them. Many kinds of monkeys delight in fondling and being fondled by each other and by persons to whom they are attached.

Two chimpanzees when first brought together sat opposite, touching each other with their much-protruded lips; and the one put his hand on the shoulder of the other. They then mutually folded each other in their arms. Afterwards they stood up, each with one arm on the shoulder of the other, lifted up their heads, opened their mouths, and yelled with delight.

We Europeans are so accustomed to kissing as a mark of affection, that it might be thought to be innate in mankind; but this is not the case. A Fuegian told me that this practice was unknown in his land. According to Sir John Lubbock it is equally unknown with the New Zealanders and Tahitians, though Wyatt Gill has seen kissing among the Papuans, Australians, Somalis and the Esquimaux. Mr. Winwood Reade says that kissing is unknown throughout West Africa, "which is probably the largest non-kissing region on the globe." But it is so far innate or natural that it apparently depends on pleasure from close contact with a beloved person; and it is replaced in various parts of the world by the rubbing of noses, as with the New Zealanders and Laplanders, by the rubbing or patting of the arms, breasts, or stomachs, or by one man striking his own face with the hands or feet of another.

The feelings which are called tender are in themselves of a pleasurable nature, excepting when pity is too deep, or horror is aroused, as in hearing of a tortured man or animal. They are remarkable from so readily exciting the secretion of tears. Many a father and son have wept on meeting after a long separation, especially if the meeting has been unexpected. No doubt extreme joy by itself tends to act on the lacrymal glands; but on such occasions as the foregoing, vague thoughts of the grief which would have been felt had the father and son never met, will probably have passed through their minds; and grief naturally leads to the secretion of tears. Thus on the return of Ulysses :—

> "Telemachus
> Rose, and clung weeping round his father's breast.
> There the pent grief rained o'er them, yearning thus.
> * * * * * *
> Thus piteously they wailed in sore unrest,
> And on their weepings had gone down the day,
> But that at last Telemachus found words to say."
>
> *Worsley's Translation of the Odyssey*,
> Book XVI, st. 27.

So, again, when Penelope at last recognized her husband:—

> "Then from her eyelids the quick tears did start,
> And she ran to him from her place, and threw
> Her arms about his neck, and a warm dew
> Of kisses poured upon him, and thus spake."
>
> Book XXIII, st. 27.

The vivid recollection of our former home, or of long-past happy days, readily causes the eyes to be suffused with tears; but here, again, the thought naturally occurs that these days will never return. In such cases we may be said to sympathize with ourselves in our present, in comparison with our former state. Sympathy with the sorrows of others, even with the imaginary distress of a heroine in a pathetic story, readily excites tears. So does sympathy with the happiness of others, as with that of a lover, at last successful after many hard trials in a well-told tale.

Sympathy, whether received or given, appears to constitute a separate or distinct emotion; and it is especially apt to excite the lacrymal glands. Every one must have noticed how readily children burst out crying if we pity them for some small hurt. A kind word will often plunge the melancholic insane into unrestrained weeping. As soon as we express our pity for the grief of a friend, tears often come into our own eyes. The feeling of sympathy is commonly explained by assuming that, when we see or hear of suffering in another, the idea of suffering is called up so vividly in our own minds that we ourselves suffer.

Sympathy with the distresses of others often excites tears more freely than our own distress. Many a man from whose eyes no suffering of his own could wring a tear has shed tears at the sufferings of a beloved friend. Again, sympathy with the happiness or good fortune of those whom we tenderly love leads to the same result, whilst a similar happiness felt by ourselves may leave our eyes dry. This is due to the fact that the long-continued habit of restraint which is so powerful in checking the free flow of tears from bodily pain, has not been brought into play in preventing a moderate effusion of tears in sympathy with the sufferings or happiness of others.

Music has a wonderful power of recalling, in a vague and indefinite manner, those strong emotions which were felt during long-past ages, when, as is probable, our early progenitors courted each other by the aid of vocal tones. And as several of our strongest emotions—grief, great joy, love, and sympathy—lead to the free secretion of tears, it is not surprising that music should be apt to cause our eyes to become suffused with tears, especially when we are already softened by any of the tenderer feelings. Music often produces another peculiar effect. We know that strong sensations, emotions, or excitements—pain, rage, terror, joy, or the passion of love—all have a special tendency to cause the muscles to tremble; and the thrill which runs down the backbone and limbs of many persons when they are powerfully affected by music, seems to bear the same relation to the above trembling of the body as a slight suffusion of tears from the power of music does to weeping from any strong and real emotion.

Devotion.—Devotion is, in some degree, related to affection, though it mainly consists of reverence, often combined with fear. With some sects, both past and present, religion and love have been strangely combined; and it has even been maintained, lamentable as the fact may be, that the holy kiss of love differs but little from that which a man bestows on a woman, or a woman on a man. As the eyes are often turned up in prayer, without the mind being so much absorbed in thought as to approach to the unconsciousness of sleep, the movement is probably a conventional one—the result of the common belief that Heaven, the source of Divine power to which we pray, is seated above us.

A humble kneeling posture, with the hands upturned and palms joined, appears to us, from long habit, a gesture so appropriate to devotion that it might be thought to be innate; but I have not met with any evidence to this effect among non-European races. During the classical period of Roman history it does not appear that the hands were thus joined during prayer. Mr. Hensleigh Wedgwood has apparently given [1] the true explanation. "When the suppliant kneels and holds up his hands with the palms joined, he represents a captive who proves the completeness of his submission by offering up his hands to be bound by the victor. It is the pictorial representation of the Latin *dare manus*, to signify submission." Hence it is not probable that either the uplifting of the eyes or the joining of the open hands, under the influence of devotional feelings, is innate or a truly expressive action; and this could hardly have been expected, for it is very doubtful whether feelings, such as we should now rank as devotional, affected the hearts of uncivilized men during past ages.

[1] "The Origin of Language." Mr. Tylor (*Early History of Mankind*) gives a more complex origin to the position of the hands during prayer.

CHAPTER IX

REFLECTION—MEDITATION—ILL-TEMPER—SULKINESS—DETERMINATION

THE corrugators, by their contraction, lower the eyebrows and bring them together, producing the vertical furrows of a frown on the forehead (Fig. 22). Sir C. Bell, who erroneously thought that the corrugator was peculiar to man, ranks it as "the most remarkable muscle of the human face. It knits the eyebrows with an energetic effort, which unaccountably, but irresistibly, conveys the idea of mind . . . when the eyebrows are knit, energy of mind is apparent, and there is the mingling of thought and emotion with the savage and brutal rage of the mere animal."

The corrugators have become much more developed in man than in the anthropoid apes; for they are brought into incessant action by him under various circumstances, and will have been strengthened and modified by the inherited effects of use. Together with the orbiculars, they protect the eyes from being too much gorged with blood during violent expiratory movements. When the eyes are closed quickly and forcibly to save them from being injured by a blow, the corrugators contract. With savages or other men whose heads are uncovered, the eyebrows are continually lowered and contracted to serve as a shade against a too strong light; and this is effected partly by the corrugators. This movement would have been more especially serviceable to man as soon as his early progenitors held their heads erect.

A man may be absorbed in the deepest thought and his brow will remain smooth until he encounters some obstacle in his train of reasoning, or is interrupted by some disturbance, and then a frown passes like a shadow over his brow. I have noticed that almost every one instantly frowns if he perceives a bad taste in what he is eating. I asked several persons, without explaining my object, to listen intently to a very gentle tapping sound, the nature and source of which they all perfectly knew, and not one frowned; but a man who joined us, and who could not conceive what we were all doing in profound silence, when asked to listen, frowned and said he could not in the least understand what we all wanted. Dr. Piderit says that stammerers generally frown in speaking; and that a man in doing even so trifling a thing as pulling on a tight boot frowns. Some persons are such habitual frowners that the mere effort of speaking almost always causes their brows to contract.

Men of all races frown when perplexed in thought, though not when absorbed only.

We may conclude, then, that frowning is not the expression of reflection, or of attention, but of something difficult or displeasing encountered in a train of thought. Deep reflection can, however, seldom be long carried on without some difficulty, so that it will generally be accompanied by a frown. Hence it is that frowning commonly gives to the countenance an aspect of intellectual energy.

The earliest and almost sole expression seen during the first days of infancy, and then often exhibited, is that of screaming; and screaming is excited by displeasing sensations and emotions. At such times the muscles round the eyes are strongly contracted; and this, as I believe, explains to a large extent the act of frowning during the remainder of our lives.

As the habit of contracting the brows at the commencement of crying or screaming has been followed by infants during innumerable generations, it has become firmly associated with the sense of something disagreeable. Hence under similar circumstances it would be apt to be continued during maturity, although never then developed into a crying-fit. Screaming or weeping is voluntarily restrained at an early period of life, whereas frowning is hardly ever restrained at any age. It is not more surprising that the habit of contracting the brows at the first perception during infancy of something distressing, should be retained during the rest of our lives, than that many other associated habits acquired at an early age should be retained throughout life. For instance, full-grown cats, when feeling warm and comfortable, retain the habit of alternately protruding their fore-feet with extended toes, which habit they practised for a definite purpose whilst sucking their mothers.

Another and distinct factor has probably strengthened the habit of frowning, whenever the mind is intent on any subject and encounters some difficulty. Vision is the most important of all the senses, and during primeval times the closest attention must have been incessantly directed towards distant objects for the sake of obtaining prey or avoiding danger. I remember being struck, whilst travelling in parts of South America which were dangerous from the presence of Indians, how incessantly, yet as it appeared unconsciously, the half-wild Gauchos closely scanned the whole horizon. Now, when any one with no covering on his head (as must have been aboriginally the case with mankind) strives to the utmost to distinguish in broad daylight a distant object, he almost invariably contracts his brows to prevent the entrance of too much light; the lower eyelids, cheeks, and upper lip being at the same time raised, so as to lessen the orifice of the eyes. I have purposely asked several persons, young and old, to look, under the above circumstances, at distant objects, making them believe that I only wished to test the power of their vision; and they all behaved in the manner just described. Some of them, also, put their open, flat hands over their eyes to keep out the excess of light. Mr. Henry Reeks writes: "I have seen the black bear, *U. americanus*, sit on its haunches and shade its eyes with both fore-paws when trying to make out a distant object, and I hear that it is a frequent habit of that species." The muscles round the eyes contract partly for the sake of excluding too much light and partly to allow only of rays striking the retina which come direct from the object scrutinized. The Rev. H. H. Blair, Principal of Worcester College, states that the born-blind have little or no control over the *corrugator supercilii*, so that they cannot frown when told to do so; nevertheless, they frown involuntarily. They can, however, smile at command.

As the effort of viewing a distant object with care under a bright light is both difficult and irksome, and as this effort has been habitually accompanied during numberless generations by the contraction of the eyebrows, the habit of frowning will thus have been much strengthened; although it was originally practised during infancy from a quite independent cause—namely, as the first step in the protection of the eyes during screaming. There is, indeed, much analogy, as far as the state of the mind is concerned, between intently scrutinizing a distant object, and following out an obscure train of thought, or performing some little and troublesome mechanical work. The belief that the habit of contracting the brows is continued when there is no need whatever to exclude too much light, receives support from the cases formerly alluded to, in which the eyebrows or eyelids are acted on under certain circumstances in a useless manner, from having been similarly used under analogous circumstances, for a serviceable purpose. For instance, we voluntarily close our eyes when we do not wish to see any objects, and we are apt to close them when we reject a proposition as if we could not or would not see it, or when we think about something horrible. We raise our eyebrows when we wish to see quickly all round us, and we often do the same when we earnestly desire to remember something; acting as if we endeavoured to see it.

Abstraction.—When a person is "lost in thought" or "in a brown study," he does not frown, but his eyes appear vacant. The lower eyelids are generally raised and wrinkled, in the same manner as when a short-sighted person tries to distinguish a distant object; and the upper orbicular muscles are at the same time slightly contracted.

The eyes are not fixed on any object, and hence the lines of vision often become slightly divergent.

Perplexity is often accompanied by certain gestures, such as raising our hands to our foreheads, mouths, or chins; but we do not act thus when quite lost in meditation. Plautus, describing a puzzled man, says, "He has pillared his chin upon his hand." Even so trifling and apparently unmeaning a gesture as the raising of the hand to the face has been observed with some savages. Kaffirs when puzzled have been described as "pulling their beards." Indians in the western regions of the United States, when concentrating their thoughts, bring the thumb and index finger in contact with the upper lip.

Ill-temper.—Frowning, with some depression of the corners of the mouth, gives an air of peevishness. If a child frowns much whilst crying, but does not strongly contract the orbicular muscles, a well-marked expression of anger or even of rage together with misery, is displayed.

If the whole frowning brow be drawn much downward by the contraction of the pyramidal muscles of the nose, which produces transverse wrinkles or folds across the base of the nose, the expression becomes one of moroseness.

A firmly closed mouth, in addition to a lowered and frowning brow, gives determination to the expression, and makes it sullen. An expression of sullen obstinacy has been clearly recognized in the natives of six different regions of Australia, in the Hindoos, Malays, Chinese, Kaffirs, Abyssinians, the Araucanos of Southern Chili, and in a conspicuous degree with the wild Indians of North America, and with the Aymaras of Bolivia. The natives of Australia, when in this frame of mind, sometimes fold their arms across their breasts, an attitude which may be seen with us. Firm determination is also sometimes expressed by both shoulders being kept raised.

With young children sulkiness and, to a lesser extent, shyness, is shown by pouting, an action that consists of the protrusion of both lips into a tubular form, sometimes to such an extent as to project as far as the end of the nose. Pouting is generally accompanied by frowning, and sometimes by the utterance of a booing or whooing noise. This expression is almost the sole one which is exhibited much more plainly during childhood than during maturity. There is, however, some tendency to the protrusion of the lips with adults under the influence of great rage.

Pouting is common and strongly marked with most savage races. This movement apparently results from the retention, chiefly during youth, of a primordial habit, or from an occasional reversion to it. Young orangs and chimpanzees protrude their lips to an extraordinary degree when they are discontented, angry, sulky, surprised, frightened, and even when pleased (Fig. 14). The mouth is protruded apparently for the sake of making the various noises proper to these several states of mind; and its shape, as I observed with the chimpanzee, differed slightly when the cry of pleasure and that of anger were uttered. As soon as these animals become enraged, the shape of the mouth wholly changes, and the teeth are exposed. The adult orang when wounded is said to emit "a singular cry, consisting at first of high notes which at length deepen into a low roar. While giving out the high notes he thrusts out his lips into a funnel shape, but in uttering the low notes he holds his mouth wide open."[1] With the gorilla the lower lip is said to be capable of great elongation.

[1] Müller, as quoted by Huxley, *Man's Place in Nature*.

If our semi-human progenitors protruded their lips when sulky or a little angered in the same manner as do the existing anthropoid apes, it is not an anomalous fact that our children should exhibit, when similarly affected, a trace of the same expression, together with some tendency to utter a noise. For animals are prone to retain during early youth, and subsequently to lose, characters which were aboriginally possessed by their adult progenitors.

Nor is it an anomalous fact that the children of savages should exhibit a stronger tendency to protrude their lips, when sulky, than the children of civilized Europeans; for the essence of savagery seems to consist in the retention of a primordial condition. It may be objected to this view of the origin of pouting, that the anthropoid apes protrude their lips both when astonished and when pleased; whilst with us this expression is generally confined to a sulky frame of mind. But with men of various races moderate surprise does sometimes lead to slight protrusion of the lips, though great surprise is shown by the mouth being widely opened.

A little gesture made by sulky children is that of "showing a cold shoulder." This has a different meaning, I believe, from the keeping both shoulders raised. A cross child, sitting on its parent's knees, will lift up the near shoulder, then jerk it away as if from a caress, and afterwards give a backward push with it as if to push away the offender. I have seen a child, standing at some distance from any one, clearly express antipathy by raising one shoulder, giving it a little backward movement, and then turning away its whole body.

Determination.—The firm closure of the mouth gives an expression of determination or decision to the countenance. No man with a gaping mouth could look determined. Hence, also, a small and weak lower jaw, which seems to indicate that the mouth is not habitually and firmly closed, is characteristic of feebleness of character. A prolonged effort of any kind, whether of body or mind, implies previous determination; and if the mouth is generally closed with firmness before and during exertion, then, through the principle of association, it would almost certainly be closed as soon as any determined resolution was taken.

When two men are engaged in a deadly contest a terrible silence prevails, broken only by hard stifled breathing. There is silence, because to expel the air in the utterance of sound would be to relax the support for the muscles of the arms. If an outcry occurs, one of the two has given up in despair.

It may be that the firm closure of the mouth during strong muscular exertion is also dependent on the principle that the influence of the will spreads to other muscles besides those necessarily brought into action in making any particular exertion. This perhaps explains why we press the teeth hard together during violent exertion.

Lastly, when a man has to perform some delicate and difficult operation not requiring violent exertion, he nevertheless generally closes his mouth and holds his breath; but he acts thus in order that his chest movements may not disturb those of his arms. A person, for instance, whilst effecting some particularly delicate manipulation compresses his lips and breathes as quietly as possible. So it was with the little sick chimpanzee whilst it amused itself by killing flies with its knuckles. To perform a difficult action, however trifling, implies some amount of previous determination. The habitual and firm closure of the mouth would thus come to show decision of character.

CHAPTER X

HATRED AND ANGER

IF we have suffered or expect to suffer some wilful injury from a man, or if he is in any way offensive to us, we dislike him; and dislike easily rises into hatred. Such feelings, if experienced in a moderate degree, are not clearly expressed by any movement of the body or features, excepting perhaps by a certain gravity of behaviour or by some ill-temper. Few individuals, however, can long reflect about a hated person without feeling and exhibiting indignation. But if the offending person be quite insignificant we experience merely disdain or contempt. If, on the other hand, he is all-powerful, then hatred passes into terror, as when a slave thinks about a cruel master, or a savage about a bloodthirsty malignant deity. Most of our emotions are so closely connected with their expression that they hardly exist if the body remains passive—the nature of the expression depending in chief part on the nature of the actions which have been habitually performed under this particular state of the mind. A man, for instance, may know that his life is in extreme peril, and may strongly desire to save it, yet may exclaim, as did Louis XVI when surrounded by a fierce mob, "Am I afraid? Feel my pulse."[1] So a man may intensely hate another, but until his bodily frame is affected he cannot be said to be enraged.

Rage.—During rage the heart and circulation are affected; the face reddens and the veins on the forehead and neck are distended. The reddening of the skin has been observed even in copper-coloured Indians, in the

[1] Compare the popular slogan during the black years of the Great War: "Are we down-hearted? *No!*"—C. M. B.

dark chocolate-coloured Papuans, and on the white cicatrices left by old wounds on negroes. Dr. Burgess (*Physiology of Blushing*) speaks of the reddening of a cicatrix in a negress as of the nature of a blush. Monkeys also redden from passion. With one of my own infants, under four months old, the first symptom of approaching passion was the rushing of blood into his bare scalp. On the other hand, the action of the heart is sometimes so much impeded by great rage that the countenance becomes pallid or livid, and not a few men with heart-disease have dropped down dead under this powerful emotion.

The respiration is likewise affected; the chest heaves and the dilated nostrils quiver. Asthmatic patients acquire permanently expanded nostrils, owing to the habitual contraction of the elevatory muscles of the wings of the nose, to allow free breathing whilst the mouth is closed and the teeth clenched. The nostrils of an angry man also become dilated although his mouth is open. Tennyson writes, "Sharp breaths of anger puffed her fairy nostrils out." Hence we have such expressions as "breathing out vengeance" and "fuming with anger." Mr. Wedgwood, in his *On the Origin of Language,* observes that the sound of hard breathing "is represented by the syllables *puff, huff, whiff,* whence a *huff* is a fit of ill-temper."

The excited brain gives strength to the muscles and energy to the will. The body is commonly held erect ready for instant action, but sometimes it is bent forward towards the offending person, with the limbs more or less rigid. The mouth is generally closed with firmness, showing fixed determination, and the teeth are clenched or ground together. Such gestures as the raising of the arms, with the fists clenched as if to strike the offender, are common. Few men in a great passion and telling some one to be gone can resist acting as if they intended to strike or push the man violently away. The desire, indeed, to strike often becomes so intolerably strong that inanimate objects are struck or dashed to the ground; but the gestures frequently become altogether purposeless or frantic. Young children when in a violent rage roll on the ground on their backs or bellies, screaming, kicking, scratching, or biting everything within reach. So it is with Hindoo children, and with the young of the anthropomorphous apes.

But the muscular system is often affected in a wholly different way; for trembling is a frequent consequence of extreme rage. The paralyzed lips then refuse to obey the will, and the voice "sticks in the throat" or is rendered loud and discordant. If there be much and rapid speaking, the mouth may froth and the hair sometimes bristles. There is in most cases a strongly marked frown on the forehead; for this follows from the sense of anything displeasing or difficult, together with concentration of mind. But sometimes the brow, instead of being contracted and lowered, remains smooth, with the glaring eyes kept widely open. The eyes are always bright or may, as Homer expresses it, be like a blazing fire. They are sometimes bloodshot, and are said to protrude from their sockets—the result, no doubt, of the head being gorged with blood, as shown by the veins being distended. The pupils are always contracted in rage.

Shakespeare sums up the chief characteristics of rage as follows :—

> "When the blast of war blows in our ears,
> Then imitate the action of the tiger :
> Stiffen the sinews, summon up the blood,
> Then lend the eye a terrible aspect;
> Now set the teeth, and stretch the nostril wide,
> Hold hard the breath, and bend up every spirit
> To his full height! On, on, you noblest English."
>
> *Henry V*, Act III, sc. 1.

The lips, not only with Europeans, but also with Australians[1] and Hindoos, are sometimes protruded during rage. More commonly, however, they are

[1] Whenever Darwin uses the words Australians, Africans, and New Zealanders, he is referring to the aborigines of the countries.—C. M. B.

retracted, the grinning or clenched teeth being thus exposed. It is possible that such action is a remnant of a habit acquired during primeval times when our semi-human progenitors fought together with their teeth, like gorillas and orangs at the present day. The general appearance of rage is as if the teeth were uncovered, ready for seizing or tearing an enemy, though there may be no intention of acting in this manner. Dr. Piderit speaks of the retraction of the upper lip during rage. In an engraving of one of Hogarth's pictures passion is represented in the plainest manner by the open glaring eyes, frowning forehead, and exposed grinning teeth. Dr. Comrie describes the natives of New Guinea as showing their canine teeth and spitting when angry. This grinning expression has been observed in both Australians and Kaffirs when quarrelling. Dickens, in speaking of an atrocious murderer who had just been caught and was surrounded by a furious mob, describes "the people as jumping up one behind another, snarling with their teeth, and making at him like wild beasts." Children naturally take to biting when in a passion. It seems as instinctive in them as in young crocodiles, who snap their little jaws as soon as they emerge from the egg. The retraction of the lips and uncovering of the teeth during paroxysms of rage, as if to bite the offender, is especially noticeable with the insane.

Dr. Maudsley asks whether such traits as the baring of the teeth in rage are not due to the reappearance of primitive instincts. Since every human brain passes, in the course of its development, through the same stages as those occurring in the lower vertebrate animals, and as the brain of an idiot is in an arrested condition, we may presume that it "will manifest its most primitive functions, and no higher functions. . . . Whence come the savage snarl, the destructive disposition . . . the wild howl . . . displayed by some of the insane? Why should a human being, deprived of his reason, ever become so brutal in character, as some do, unless he has the brute nature within him?" Dr. Maudsley's question must, it would appear, be answered in the affirmative.

Anger.—This state of the mind differs from rage only in degree. Under moderate anger the action of the heart is a little increased, the colour heightened, and the eyes become bright. The respiration is a little hurried, and the wings of the nostrils are somewhat raised to allow of a free indraught of air. The mouth is commonly compressed, and there is almost always a frown. Instead of the frantic gestures of extreme rage, an indignant man unconsciously throws himself into an attitude ready for attacking his enemy, whom he will perhaps scan from head to foot in defiance. He carries his head erect with his chest well expanded. He holds his arms with one or both elbows squared or rigidly suspended by his sides. With Europeans the fists are commonly clenched. Clenching the fists, however, seems confined chiefly to men who fight with their fists. Only one of my informants has seen this gesture among Australians. With these people during anger and rage the body is held erect, the brows are heavily contracted, the mouth firmly compressed, the nostrils are distended, the eyes flash and are widely opened, while the lips are protruded. The men wave their arms, while the women dance about and cast dust into the air.

Abyssinians, the natives of South Africa, and the Dakota Indians of North America behave not dissimilarly, and the last-named will hold their heads erect, frown, and often stalk away with long strides. Fuegians, when enraged, frequently stamp on the ground, walk distractedly about, sometimes cry and grow pale. The Rev. Mr. Stack thus describes a New Zealand man and woman quarrelling: "Eyes dilated, body swayed violently backwards and forwards, head inclined forwards, fists clenched, now thrown behind the body, now directed towards each other's faces."

Two low-caste Bengalees disputed about a loan. At first they were calm, but soon grew furious and poured forth the grossest abuse on each other's relations for many generations past. Their chests were expanded and shoulders squared, their arms remained rigidly suspended with the elbows turned inwards and the hands alternately clenched and opened. Their shoulders were often raised high and then again lowered. They looked fiercely at each other from under their lowered and strongly wrinkled brows, and their protruded lips were firmly closed. They approached each other, with heads and necks stretched forwards, and pushed, scratched, and grasped at each other. This protrusion of the head and body seems a common gesture with the enraged; and I have noticed it with Englishwomen whilst quarrelling. The protruding of the head and body towards the offender by an enraged man may be a remnant of attacking an enemy with the teeth, for it would appear that neither party expects to receive a blow from the other. H. N. Moseley describes an Admiralty Islander in a "furious rage"; the head is "lowered and jerked towards the object of his wrath as if he meant to attack him with his teeth."

Two Mechis in Sikhim were seen in a furious passion; their bodies became less erect, with their heads pushed forwards; they made grimaces at each other, their shoulders were raised, their arms rigidly bent inwards at the elbows, and their hands spasmodically closed and opened. They continually approached and retreated from each other, and often raised their arms as if to strike, but their hands were open and no blow was given.

Sneering.—This expression differs from that already described, solely in the upper lip being retracted in such a manner that the canine tooth on one side alone is shown, the face being generally a little upturned and half averted from the person causing offence (Fig. 18). This expression may occasionally be observed in a person who sneers at or defies another, though there may be no real anger; as when any one is playfully accused of some fault and answers, "I scorn the imputation." Parsons, as long ago as 1746, depicted scorn with an engraving, showing the uncovered canine on one side.

The expression of a half-playful sneer graduates into one of great ferocity when, together with a heavily frowning brow and fierce eye, the canine tooth is exposed. A Bengalee boy was accused of some misdeed. The delinquent plainly showed his wrath on his countenance, sometimes by a defiant frown and sometimes "by a thoroughly canine snarl." When

FIG. 18.—Sneering and defiance; the canine tooth on one side is uncovered.

this was exhibited "the corner of the lip over the eye-tooth, which happened in this case to be large and projecting, was raised on the side of his accuser, a strong frown being still retained on the brow." The actor Cooke could express the most determined hate when he drew up the outer part of the upper lip and exposed a sharp angular tooth. Sir C. Bell calls the muscles which uncover the canines the *snarling muscles.*

The action of uncovering the canine tooth is the same as that of a snarling dog (Fig. 11); and a dog when pretending to fight often draws up the lip on one side alone—namely, that facing his antagonist. Our word *sneer* is, in fact, the same as *snarl*, which was originally *snar*.

I suspect there is a trace of this same expression in the derisive or sardonic smile. The lips are then kept joined or almost joined, but one corner of the mouth is retracted on the side towards the derided person; and this drawing back of the corner is part of a true sneer. I have also on these occasions noticed a slight twitching of the muscle which draws up the outer part of the upper lip; and this movement, if fully carried out, would have uncovered the canine and would have produced a true sneer.

Mr. Bulmer, an Australian missionary in a remote part of Gipps' Land, says: "I find that the natives in snarling at each other speak with the teeth closed, the upper lip drawn to one side, and a general angry expression of face; but they look direct at the person addressed." It is probable that this animal-like expression may be more common with savages than with civilized races.

It is surprising that man should possess the power of uncovering his canine tooth, for a snarling action is not noticeable in our nearest allies, the monkeys. Baboons, though furnished with great canines, uncover all their teeth when feeling savage and ready for an attack. Whether the adult anthropomorphous apes, in the males of whom the canines are much larger than in the females, uncover them when prepared to fight, I do not know.

The expression, whether that of a playful sneer or ferocious snarl, is one of the most curious which occurs in man. It reveals his animal descent; for no one, even if rolling on the ground in a deadly grapple with an enemy and attempting to bite him, would try to use his canine teeth more than his other teeth. We may readily believe from our affinity to the anthropomorphous apes that our male semi-human progenitors possessed great canine teeth, and occasionally men are now born having them of unusually large size with interspaces in the opposite jaw for their reception. We may further suspect that our semi-human progenitors uncovered their canine teeth when prepared for battle, as we still do when feeling ferocious, or when merely sneering at or defying some one.

Chapter XI

DISDAIN—CONTEMPT—DISGUST—GUILT—PRIDE, ETC.—HELPLESSNESS—PATIENCE—AFFIRMATION AND NEGATION

Disgust—a combination of emotion and sensation—is distinct in its nature, and refers to something revolting, primarily in relation to the sense of taste, as actually perceived or vividly imagined; and secondarily to anything which causes a similar feeling, through the sense of smell, touch, or vision. Nevertheless, extreme contempt hardly differs from disgust.

The derisive smile implies that the offender is so insignificant that he excites only amusement; but the amusement is generally a pretence. Contempt is commonly shown by Kaffirs and the Dyaks of Borneo by smiling. As laughter is primarily the expression of simple joy, very young children never laugh in derision.

Partial closure of the eyelids and the turning away of the eyes or of the whole body are likewise highly expressive of disdain. These actions seem to declare that the despised person is not worth looking at, or is disagreeable to behold. The accompanying photograph (Fig. 19) by Mr. Rejlander shows this form of disdain. It represents a young lady who is supposed to be tearing up the photograph of a despised lover. Mr. H. Holbeach suggests that when "the head is lifted upwards and backwards in order to give the feeling of as much distance in the way of height as possible being placed between the despiser and the despised, the eyelids partake of the general movement, and the eyes are made to look *down* upon the object of contempt."

Professor Cleland writes: "In haughtiness the

upward head contrasts with the somewhat downward glance, indicating that it is the height pertaining to self which occupies the mind and which looks down on others."

The most common method of expressing contempt is by movements about the nose or round the mouth; but the latter movements, when strongly pronounced, indicate disgust. The nose may be slightly turned

FIG. 19.—The expression is that of disdain; the eyes "look down upon" the object of contempt.

up, which apparently follows from the turning up of the upper lip; or the movement may be abbreviated into the mere wrinkling of the nose. The nose is often slightly contracted, so as partly to close the passage; and this is commonly accompanied by a slight snort or nasal expiration. All these actions are the same with those which we employ when we perceive an offensive odour and wish to exclude or expel it. When we wish to smell carefully we draw in the air by a succession of rapid short sniffs. The nostrils, far from dilating, actually contract at each sniff. When we wish to *exclude* any odour the contraction, I presume, affects only the anterior part of the nostrils. In extreme cases of contempt and disgust we protrude and raise both lips, or the upper lip alone, so as to close the nostrils as by a valve; the nose being thus turned up. It is as though we told the despised person that he smells offensively, in nearly the same manner as we express to him, by half closing our eyelids or turning away our faces, that he is not worth looking at. One of the roots of the word "scorn" means ordure or dirt. A person who is scorned is treated like dirt. It must not, however, be supposed that such ideas actually pass through the mind when we exhibit contempt; but as whenever we experience a disagreeable odour or sight actions of this kind have been performed, they have become fixed, and are now employed under any analogous state of mind.

Various odd little gestures likewise indicate contempt; for instance, *snapping one's fingers*. Strabo makes mention of this gesture. There is an Eastern equivalent of snapping the fingers to express contempt and imply minuteness, nothing, or negation—namely, touching the upper front teeth with the thumb nail and then snapping it away, as if throwing away a bit of the nail.

The phrase, "Do you bite your thumb at us, sir?", in *Romeo and Juliet*, refers to a similar gesture of contempt. With the Dakota Indians contempt is shown not only by movements of the face such as those above described, but by the hand being closed and held near the breast; then, as the forearm is suddenly extended, the hand is opened and the fingers separated from each other. The hand is moved towards the offending person and the head sometimes averted from him. This sudden extension and opening of the hand perhaps indicates the dropping or throwing away a valueless object.

The term "disgust," in its simplest sense, means something offensive to the taste. It is curious how readily this feeling is excited by anything unusual in the appearance, odour, or nature of our food. In

Tierra del Fuego a native touched some cold meat which I was eating, and plainly showed utter disgust at its softness; whilst I felt utter disgust at my food

FIG. 20.—Disgust. The expression is assumed chiefly by movements around the mouth accompanied by the gesture of pushing away or guarding oneself against the offensive object.

being touched by a naked savage, though his hands did not appear dirty. A smear of soup on a man's beard looks disgusting, though there is, of course, nothing disgusting in the soup itself. I presume that this follows from the strong association in our minds between the sight of food, however circumstanced, and the idea of eating it.

As the sensation of disgust primarily arises in connection with the act of eating or tasting, it is natural that its expression should consist chiefly in movements round the mouth. But as disgust also causes annoyance it is generally accompanied by a frown, and often by gestures as if to push away or to guard oneself against the offensive object. The photograph (Fig. 20) well exhibits this expression of disgust. Moderate disgust is shown in various ways: by the mouth being widely opened, as if to let an offensive morsel drop out; by spitting; by blowing out of the protruded lips; or by a sound as of clearing the throat. Such guttural sounds are written *ach* or *ugh*; and their utterance is sometimes accompanied by a shudder, the arms being pressed close to the sides and the shoulders raised in the same manner as when horror is experienced. Extreme disgust is expressed by movements round the mouth identical with those preparatory to the act of vomiting.

Retching or actual vomiting is readily and instantly induced in some persons by the mere idea of having partaken of unusual food, as of an animal which is not commonly eaten; although there is nothing in such food to cause the stomach to reject it. When vomiting results as a reflex action from too rich food, or tainted meat, or from an emetic—it does not ensue immediately, but generally after a considerable interval of time. Therefore, to account for retching or vomiting being so quickly and easily excited by a mere idea, the suspicion arises that our progenitors must have had the power like that possessed by ruminants and some other animals of voluntarily rejecting food which disagreed with them or which they thought would disagree with them; and now, though voluntary power has been lost, it is called into involuntary action, through the force of a formerly well-established habit, whenever the mind revolts at the idea of having partaken of any kind of food or at anything disgusting. This suspicion receives further support from the fact that monkeys often vomit whilst in perfect health, which looks as if the act were voluntary. There is on record the case of an idiot who had the power of voluntarily regurgitating food from the stomach; also of a Scotch youth who had a similar power of bringing up his food.

Bitches often vomit up food for their young when these have reached a certain age. As man acquired the ability to communicate by language to his children knowledge of food to be avoided, occasion to use the faculty of voluntary rejection would gradually lapse, so that this power would tend to be lost through disuse.

As the sense of smell is intimately connected with that of taste, it is not surprising that an excessively bad odour should also excite retching or vomiting in some persons, and that, as a further consequence, a moderately offensive odour should cause the various expressive movements of disgust. On a certain occasion I wished to clean the skeleton of a bird which had not been sufficiently macerated, and the smell made my servant and myself retch so violently that we were compelled to desist. During the previous days I had examined some other skeletons which smelt slightly; yet the odour did not in the least affect me, but subsequently for several days whenever I handled these same skeletons they made me retch.

A Greenlander expresses contempt or horror by turning up his nose and giving a slight sound through it. Mr. Bridges says that the Fuegians "express contempt by shooting out the lips and hissing through them, and by turning up the nose." The inhabitants of New Guinea express disgust by pouting or by an imitation of vomiting. The tendency either to snort through the nose or to make a noise resembling *ugh* or *ach* is common as an expression of disgust.

Spitting seems an almost universal sign of contempt or disgust; and spitting obviously represents the rejection of anything offensive from the mouth. Shakespeare makes the Duke of Norfolk say, "I spit at him—call him a slanderous coward and a villain." So, again, Falstaff says, " Hal, if I tell thee a lie, spit in my face." Native Australians will interrupt their speeches by spitting and uttering a noise like pooh! pooh! apparently expressive of their disgust, and Abyssinians and negroes are known to spit with disgust upon the ground. With the Malays of Malacca disgust is expressed by spitting, and among the Fuegians, to spit at a person is the highest mark of contempt.

Scorn, disdain, contempt, and disgust are the same throughout the world, and all consist of actions representing the rejection or exclusion of some object which we dislike or abhor.

Jealousy, Suspicion, Slyness, Guilt, Vanity, Pride, Humility, etc.—Difficult though such complex mental states are to describe, they are, nevertheless, readily detected by the eye. Shakespeare speaks of Envy as *lean-faced*, or *black*, or *pale*, and Jealousy as "*the green-eyed monster*"; and Spenser describes Suspicion as "*foul, ill-favoured, and grim*" (Fig. 7). The guilty man is said to avoid looking at his accuser, or to give him "stolen looks." The eyes are said " to be turned askant," or " to waver from side to side," or " the eyelids to be lowered and partly closed." The natives of India are able to control the expression of their faces so that no indication is given as to whether they are speaking the truth or not; but they cannot control the toes, the contortions of which often reveal the fact that the witness is lying. The restless movements of the eyes apparently follow from the guilty man not enduring to meet the gaze of his accuser. Concealment or deceit is expressed by the face being directed downwards, while the eyes are turned upwards. The culprit, so to speak, shelters himself by a lie, " hangs his head over his secret, while he steals upward glances to see the effect which he distrusts." I have observed a guilty expression in my own children at a very early age; in one instance it was unmistakably clear in a child two years and seven months old, and led to the detection of his little crime. It was shown by an unnatural brightness in the eyes and by an odd, affected manner. Slyness is exhibited chiefly by movements about the eyes. When these are turned to one side, while the face is not turned to the same side, we get the expression of slyness (Fig. 17).

Pride is most plainly expressed. A proud man exhibits his sense of superiority by holding his head and body erect. He is haughty (*haut*), or high, and makes himself appear as large as possible; so that he is said to be swollen or puffed up with pride. A peacock or a turkey-cock strutting about with puffed-up feathers is sometimes said to be an emblem of pride. The arrogant man looks down on others, and with lowered eyelids hardly condescends to see them. The lower lip is at the same time often everted, and hence the muscle causing the action has been called the *musculus superbus*. The whole expression of pride stands in direct antithesis to that of humility.

Helplessness.—Bulwer, in 1649, described shrugging the shoulders as follows: " They who like not a thing that has happened . . . (or) who . . . would frame an excuse, are wont to shrink the head and contracted neck between the shoulders." When a man wishes to show that he cannot do something or prevent something being done, he often raises both shoulders with a quick movement. At the same time, if the whole gesture is completed he bends his elbows closely inwards, raises his open hands, turning them outwards, with the fingers separated. The head, as a rule with the mouth open, is often thrown a little on one side; the eyebrows are elevated, and the forehead is wrinkled.

Englishmen are less demonstrative and shrug their shoulders less frequently and energetically than most other Europeans. The gesture varies from the complex movement just described to a momentary and scarcely perceptible raising of both shoulders, or to the mere turning slightly outwards of the open hands with separated fingers. I have never seen very young English children shrug their shoulders, though the children of the French do so. The blind and deaf Laura Bridgman, who could not have learnt the habit by imitation, shrugs her shoulders, turns in her elbows, and raises her eyebrows in the same manner as other people, and under the same circumstances. The fact that the gesture has been reported as occurring among Europeans, Hindoos, the hill-tribes of India, Malays, Micronesians, Abyssinians, Arabs, Negroes, Indians of North America, and apparently among the Australians, is sufficient to show that shrugging the shoulders is a gesture natural to mankind.

The action implies an incapacity on our own part to do something or to prevent some action on the part of another person. It accompanies such words as, "It was not my fault"; "He must follow his own course, I cannot stop him." Shrugging the shoulders likewise expresses patience or the absence of any intention to resist. Hence the muscles which raise the shoulders are sometimes called "the patience muscles." Shylock the Jew says,

> "Signor Antonio, many a time and oft
> In the Rialto you have rated me
> About my moneys and my usances;
> Still have I borne it with a patient shrug."

The movement also may express a dogged determination not to act. Sulky and obstinate children often raise both their shoulders high up.

Why do men in all parts of the world when they feel that they are powerless shrug their shoulders, at the same time often bending in their elbows, showing the palms of their hands with extended fingers, often throwing their heads a little on one side, raising their eyebrows, and opening their mouths? None of the movements is of the least service. The explanation lies in the principle of unconscious antithesis, which here seems to come into play as in the case of a dog who, when feeling savage, puts himself in the attitude for attacking and for appearing terrible to his enemy; but as soon as he feels affectionate throws his whole body into a directly opposite attitude, though this is of no direct use to him. On the other hand, M. Baudry has suggested that shrugging of the shoulders is not to be explained by the principle of antithesis, that it is the natural gesture of one who receives a blow without resistance. I think, however, that the shrug of a schoolboy who is being threatened with a box on the ears is distinct from the apologetic shrug. The action of shrinking from an unseen danger, as when a cricket ball is coming towards one from behind and some one shouts "Heads!" is of the same nature as the protective shrug. M. Baudry describes it as a gesture of tucking in the head and neck. A somewhat similar shrug is familiar as an expression of feeling cold. Here it is a conscious repetition of the attitude instinctively assumed to economize the heat of the body. M. Baudry also suggests that the open hands express defencelessness, as showing that the actor has no weapons.

An indignant man who resents, and will not submit to some injury, holds his head erect, squares his shoulders, and expands his chest. He often clenches his fists and puts one or both arms in the proper position for attack or defence, with the muscles of his limbs rigid. He frowns, and, being determined, closes his mouth. The actions and attitude of a helpless man are, in every one of these respects, exactly the reverse. He unconsciously contracts the muscles of his forehead which are antagonistic to those that cause a frown, and thus raises his eyebrows; at the same time he relaxes the muscles about the mouth so that the lower jaw drops. The antithesis is complete in every detail, not only in the movements of the features, but in the position of the limbs and in the attitude of the whole body.

Nodding and Shaking the Head.—The signs of affirmation and negation are expressive of our feelings; we give a nod of approval with a smile to our children. and shake our heads laterally with a frown to them when we disapprove. With infants the first act of denial consists in refusing food; and I repeatedly noticed with my own infants that they did so by withdrawing their heads laterally from the breast, or from anything offered them in a spoon. In accepting food and taking it into their mouths they incline their heads forwards. In accepting or taking food there is only a single movement forward, and a single nod implies an affirmation. On the other hand, in refusing food, especially if it be pressed on them, children frequently move their heads several times from side to side, as we do in shaking our heads in negation. Moreover, in the case of refusal the mouth is often tightly closed, so that this movement might likewise come to serve as a sign of negation.

That these signs are innate or instinctive, at least with Anglo-Saxons, is rendered highly probable by the blind and deaf Laura Bridgman "constantly accompanying her *yes* with the common affirmative nod, and her *no* with our negative shake of the head." A microcephalous idiot, so degraded that he never learned to speak, when asked whether he wished for more food or drink answered by inclining or shaking his head.

I am informed that both signs are used by the Malays, by the natives of Ceylon, the Chinese, and the negroes of the Guinea coast. Australians and Kaffirs give a nod of affirmation. In parts of Australia a negative is expressed by throwing the head a little backwards and putting out the tongue. Near Torres Straits, the natives utter a negative by holding up the right hand and turning it half round two or three times. H. N. Moseley states that the Admiralty Islanders universally express negation by striking the nose on one side with the extended forefinger, and that both the Fijians and the Admiralty Islanders express affirmation by an upward nod. The throwing back of the head with a cluck of the tongue is said to be used as a negative by the Neapolitans and Sicilians as well as by the modern Greeks and Turks, the latter people expressing *yes* as we express *no* when we shake our heads. There is, however, some obscurity on this point.[1] Mr. Sophocles, a native of Greece, and a teacher of Greek at Harvard University, says that the Turks never express affirmation by a shake of the head. He describes Turks as gravely bowing their heads in token of approval and assent, and throwing their heads back to anything to which they cannot assent. Vesalius speaks of "Cretans" as expressing negation by an upward nod.

On the other hand, Mr. Sophocles has often seen Turks and other Orientals shake their heads in anger or strong disapprobation. This gesture is a familiar one with ourselves, and there are several instances of its occurrence in the Bible. Thus: Matt. xxvii. 39, "And they that passed by reviled Him, wagging their heads"; compare also Psalms xxii. 7, and cix. 25. The Abyssinians express a negative by jerking the head to the right shoulder, together with a slight cluck, the mouth being closed; an affirmation is expressed by the head being thrown backwards and the eyebrows raised for an instant. The Tagals of the Philippine Archipelago say *yes*, by throwing the head backwards. The Dyaks of Borneo express an affirmation

[1] Mr. Sedat Zeki, of the Turkish Embassy in London, has kindly given me the following information—:

The gesture of "*Yes*," "*Agreement*," is expressed by the Turks by a forward-downward nod of the head; "*No*," "*Disagreement*," by an opposite backward-upward movement. The gesture for "*Come here*," "*Advance*," is the holding of the fully- or half-extended arm before the body on a level with the head while the hand, palm downwards, is repeatedly moved up and down as though fanning the face. "*Go away*," "*Retire*," is expressed by holding the partially extended arm at the level of, and with the palm towards, the stomach, and moving it several times to-and-fro as in sweeping crumbs off a table.

Mr. Zeki adds that a slow shaking of the head without twisting of the neck, usually accompanied by "clucks" of the tongue, denotes hesitation, amazement, indignation, and threat, the exact shade being expressed by the face and especially the eyes. This is of interest as showing how a gesture of mild emotion may gradually merge into one of anger with, possibly, violence.—C. M. B.

by raising the eyebrows, and a negation by slightly contracting them. With the Arabs on the Nile nodding in affirmation is rare, whilst shaking the head in negation is not used, and is not even understood by them. With the Esquimaux, a nod means *yes* and a wink *no.* The New Zealanders' *yes* is an elevation of the head and chin.

With the Hindoos a nod and a lateral shake are sometimes used as with us; but a negative is more commonly expressed by the head being thrown suddenly backwards with a cluck of the tongue. Affirmation is more frequently shown by the head being first thrown backwards either to the left or right, and then jerked obliquely forwards once; in negation the head is usually held nearly upright and shaken several times.

The Fuegians nod their heads vertically in affirmation, and shake them laterally in denial. The Indians of North America express affirmation by describing with the index finger a curve downwards and outwards from the body, whilst negation is expressed by moving the open hand outwards, with the palm inwards. Sometimes they raise the forefinger and then lower it and point it to the ground, or the hand is waved straight forward from the face as a sign of affirmation; while that of negation is the finger or whole hand shaken from side to side. The Italians and sometimes we English in like manner move the lifted finger from right to left in negation. A lateral shake of the index finger or of the whole hand is a common sign of negation in Japan.

If we admit that the shaking of the finger or hand from side to side is symbolic of the lateral movement of the head; and if we admit that the sudden backward movement of the head represents one of the actions of infants in refusing food, then there is much uniformity throughout the world in the signs of negation, and we can see how they originated.

With respect to nodding in affirmation the exceptions are rather more numerous.

Chapter XII

SURPRISE—ASTONISHMENT—FEAR—HORROR

Attention graduates into surprise, this into astonishment, and this into amazement. The latter frame of mind is closely akin to terror. Attention is shown by the eyebrows being slightly raised; and as this state increases into surprise they are raised to a much greater extent, with the eyes and mouth widely open. The raising of the eyebrows is necessary in order that the eyes should be opened quickly and widely, and this movement produces transverse wrinkles across the forehead. The degree to which the eyes and mouth are opened corresponds with the degree of surprise felt; but these movements must be co-ordinated, for a widely opened mouth with eyebrows only slightly raised results in a meaningless grimace. On the other hand, a person may often be seen to pretend surprise by merely raising his eyebrows.

Shakespeare says, " I saw a smith stand with open mouth swallowing a tailor's news." And again, " They seemed almost, with staring on one another, to tear the cases of their eyes; there was speech in their dumbness, language in their very gesture; they looked as they had heard of a world destroyed."

Mr. Winwood Reade is familiar with this expression among the negroes on the Guinea coast, while other observers have noted it in the Kaffirs of South Africa, the Abyssinians, Ceylonese, Chinese, Fuegians, North Americans, and New Zealanders. the Dyaks of Borneo are said to open their eyes widely when astonished, often swinging their heads to and fro and beating their breasts, and the natives of Calcutta, when suddenly surprised, open their eyes and mouths widely.

The Australian explorer, Mr. Stuart, thus describes amazement together with terror in a native. "He stood incapable of moving a limb, riveted to the spot, mouth open and eyes staring. . . . He remained motionless until our black got within a few yards of him, when, suddenly throwing down his waddies, he jumped into a mulga bush as high as he could get." He could not speak, and answered not a word to the inquiries made by the black, but, trembling from head to foot, "waved with his hand for us to be off."

That the eyebrows are raised by an instinctive impulse may be inferred from the fact that the blind act thus when astonished. As surprise is excited by something unexpected or unknown, we naturally desire, when startled, to perceive the cause as quickly as possible; and we consequently open our eyes fully, so that the field of vision may be increased and the eyeballs moved easily in any direction. The energetic lifting up of the eyebrows opens the eyes so widely that they stare, the white being exposed all round the iris. Moreover, the elevation of the eyebrows is an advantage, for as long as they are lowered they impede our vision in an upward direction. The habit of raising the eyebrows having once been gained in order to see as quickly as possible all around us, the movement would follow from the force of association whenever astonishment was felt.

With adult persons, when the eyebrows are raised, the whole forehead becomes much wrinkled in transverse lines. These wrinkles are highly characteristic of the expression of surprise or astonishment. Each eyebrow when raised becomes also more arched than it was before.

Every sudden emotion, including astonishment, quickens the action of the heart, and with it the respiration. We can breathe more quietly through the open mouth than through the nostrils. Therefore, when we wish to listen intently to any sound we breathe as quietly as possible by opening our mouths. This view receives support from the reversed case which occurs with dogs. A dog when panting breathes loudly; but if his attention be suddenly aroused he instantly pricks his ears to listen, shuts his mouth, and breathes quietly through his nostrils.

When attention is concentrated for a length of time with fixed earnestness on any object, all the organs of the body are, so to speak, neglected; and as the nervous energy is limited in amount, little is transmitted to them or to any part of the system excepting that which is being brought into energetic action. Therefore many of the muscles become relaxed, and the jaw drops from its own weight. This accounts for the open mouth of a man stupefied with amazement. Young children when only moderately surprised open their mouths widely.

There is still another cause leading to the mouth being opened when we are astonished or suddenly startled. We can draw a deep inspiration more easily through the widely open mouth than through the nostrils. Now, when we start at any sudden sound or sight, almost all the muscles of the body are involuntarily and momentarily thrown into strong action, for the sake of jumping away from, or guarding ourselves against, the danger which we habitually associate with anything unexpected. But we always unconsciously prepare ourselves for any great exertion by first taking a full inspiration, and we consequently open our mouths. If no exertion follows, and we still remain astonished, we cease for a time to breathe, or breathe as quietly as possible, in order that every sound may be distinctly heard. Mr. Wallace suggests that among our savage ancestors danger to themselves or others would often be associated with the cause of amazement, and that the open mouth may be the rudiment, as it were, of the cry of alarm or encouragement. The opening of the mouth and protrusion, of the lips when we are astonished remind us of the same movements, though in a much more strongly marked degree, in the chimpanzee and orang when similarly affected. On a quiet night some rockets were fired from the *Beagle*, in a little creek at Tahiti, to amuse the natives; and as each rocket was let off there was absolute silence, but this was invariably followed by a deep groaning *Oh*, resounding all round the bay. Mr. W. Matthews says that the North American Indians express astonishment by a groan; and the negroes on the West Coast of Africa, according to Mr. Winwood Reade, protrude their lips, and make a sound like *heigh, heigh.* If the mouth is not much opened whilst the lips are considerably protruded, a blowing, hissing, or whistling noise is produced. An Australian from the interior was taken to the theatre to see an acrobat rapidly turning head over heels; "he was greatly astonished, and protruded his lips, making a noise with his mouth as if blowing out a match." The Australians also when surprised utter the exclamation *korki*, with the mouth drawn out as if going to whistle. Europeans often whistle as a sign of surprise. The "whew" of surprise is produced by an inspiration, whereas the "prolonged whistle" is a conscious imitation of it. A Kaffir girl, on hearing of the high price of an article, raised her eyebrows and whistled just like a European.

A surprised person often raises the opened hands to the level of or even above the head. This gesture is to be seen in children. A box of toys was opened before a child one year nine months old. She immediately threw up both hands with palms forward and fingers extended on each side of her face, crying out, oh! or ah! In the "Last Supper," by Leonardo da Vinci, two of the Apostles have their hands half uplifted, clearly expressive of their astonishment.

These gestures are, I believe, explicable on the principle of antithesis. A man in an ordinary frame of mind usually keeps his arms suspended laxly by his sides, with his hands somewhat flexed and the fingers near together. To raise the arms suddenly, to open the palms flat, and to separate the fingers, are movements in complete antithesis to those preserved under an indifferent frame of mind, and they are, in consequence, unconsciously assumed by an astonished man. Mr. Wallace, however, explains the action of the hands as appropriate movements either to defend the observer's face or body or to prepare to give assistance to another person in danger. But against this view it should be noted that there is no tendency to open the mouth under such circumstances.

Another little gesture expressive of astonishment—namely, the hand being placed over the mouth, or on some part of the head—has been observed with many races of man. Professor Gomperz suggests that in the life of a savage surprise would frequently occur on occasions when silence was needful, as on the sudden appearance or sound of an animal. The placing the hand over the mouth would thus have been originally a gesture enjoining silence, which afterwards became associated with the feeling of surprise even when no need for silence existed.

In Job xxi. 5 we read: "Mark me, and be astonished, and lay your hand upon your mouth." A wild Australian, taken into a room full of official papers which surprised him greatly, cried out *cluck, cluck,* putting the back of his hand towards his lips. The Kaffirs and Fingoes express astonishment by placing the right hand upon the mouth, uttering the word *mawo,* which means "wonderful"; the Bushmen are said to put their right hands to their necks, bending their heads backwards. Mr. Winwood Reade has observed that the negroes on the West Coast, when surprised, clap their hands to their mouths, saying, "My mouth cleaves to my hands"; the Abyssinians place their right hand to the forehead, with the palm outside. The conventional sign of astonishment with the wild tribes of the western United States "is made by placing the half-closed hand over the mouth"; the hand is also pressed over the mouth by certain Indian tribes to express astonishment.

Terror.—The word "fear" seems to be derived from what is sudden and dangerous, and that of terror, or extreme fear, from the trembling of the vocal organs and body. Fear is often preceded by astonishment, and is so far akin to it that both lead to the senses of

FIG. 21.—Terror. The eyes stare, the nostrils and pupils are dilated; the strong contraction of the *Platysma*—the "muscle of fright"—wrinkles the neck and helps to depress the lower jaw and keep the mouth open. The brow is *horizontally* furrowed and the general attitude is one of flaccidity and weakness.

sight and hearing being instantly aroused. Hence the eyes and mouth are widely opened and the eyebrows raised (Fig. 21). The frightened man at first stands like a statue motionless and breathless, or crouches down as if instinctively to escape observation. Mr. A. J. Munby gives a graphic description of terror in the case of an old woman who mistook him for "the devil or a ghost." "In an instant, with a sort of galvanic jerk, she faced me, . . . and . . . rose to her full height and stood literally on the tips of her toes, and at the same moment she threw out both her arms, placing the upper-arm nearly at right angles to her body and the forearm at right angles to the upper-arm, so that the forearms were vertical. Her hands, with the palms towards me, were spread wide, the thumbs and every finger stiff and standing apart. Her head was slightly thrown back, her eyes dilated and rounded, and her mouth wide open. . . . In opening her mouth she uttered a wild and piercing scream, and . . . the moment she recovered herself somewhat, she turned and fled, still screaming. . . . For myself, I stood gazing at her and rooted to the spot : . . . her appearance was so strange that I half fancied *her* a thing 'uncanny,' being in a house so old and lonesome, and I felt my own eyes dilating and mouth opening, though I did not utter a sound until she had fled; and then I realized the oddity of the situation and ran after her to reassure her."

During fright the heart beats quickly and violently, so that it knocks against the ribs; but it is very doubtful whether it then works more efficiently than usual, for the skin instantly becomes pale, as during incipient faintness. Mosso describes the ears of rabbits as exhibiting a momentary pallor, followed by a blush when the animals are startled. That the skin is much affected under the sense of great fear we see in the marvellous and inexplicable manner in which perspiration immediately exudes from it. This exudation is all the more remarkable as the surface is then cold, and hence the term "a cold sweat": whereas the sudorific glands are properly excited into action when the surface is heated. The hairs also on the skin stand erect, and the superficial muscles shiver. In connection with the disturbed action of the heart the breathing is hurried. The salivary glands act imperfectly; the mouth becomes dry, and is often opened and shut. Mr. Bain describes a custom in India wherein suspected criminals are subjected to the ordeal of the morsel of rice. The accused is made to take a mouthful of rice, and after a little time to throw it out. If the morsel is quite dry, the party is believed to be guilty,—his own evil conscience operating to paralyze the salivating organs. Under slight fear there is a strong tendency to yawn. One of the best-marked symptoms is the trembling of all the muscles of the body; and this is often first seen in the lips. From this cause, and from the dryness of the mouth, the voice becomes husky or indistinct, or may altogether fail.

Of vague fear there is a grand description in Job: "Fear came upon me, and trembling, which made all my bones to shake. Then a spirit passed before my face; the hair of my flesh stood up."

As fear increases into terror, the heart beats wildly, or may fail to act, and faintness ensues; there is a death-like pallor; the breathing is laboured; the wings of the nostrils are widely dilated; "there is a gasping and convulsive motion of the lips, a tremor on the hollow cheek, a gulping and catching of the throat"; the protruding eyeballs are fixed on the object of terror; or they may roll restlessly from side to side. The pupils are enormously dilated. The muscles of the body may become rigid, or may be thrown into convulsive movements. The hands are alternately clenched and opened, often with a twitching movement. The arms may be protruded as if to avert some dreadful danger, or may be thrown wildly over the head. In other cases there is a sudden and uncontrollable tendency to headlong flight; and even the boldest soldiers may be seized with a sudden panic.

As fear rises to an extreme pitch the dreadful scream of terror is heard. Great beads of sweat stand on the skin. All the muscles of the body are relaxed. Utter prostration soon follows, and the mental powers fail.

Savages exhibit the same signs of fear as Europeans and other races, but being incapable of repressing these signs to the same extent, they are apt to tremble greatly, and their sphincter muscles often relax as in terrified dogs and monkeys.

The Erection of the Hair.—Poets continually speak of the hair standing on end; Brutus says to the ghost of Cæsar, "That mak'st my blood cold, and my hair to stare." And Cardinal Beaufort, after the murder of Gloucester, exclaims, "Comb down his hair; look, look, it stands upright." The hair of a Caucasian lady became erect without the stimulus of any strong emotion. When she was affected by strong emotion her hair stirred and rose "as if it were alive," so that she herself was frightened at it.

The fact of the hair becoming erect under the influence of rage and fear agrees perfectly with what takes places in the lower animals. Bristling of the hair is frequently seen in maniacs, especially during their paroxysms of violence. In one man, prior to a maniacal paroxysm, "the hair rises up from his forehead like the mane of a Shetland pony." In mental patients in whom the bristling of the hair is extreme the disease is generally permanent; in others in whom it is moderate, as soon as they recover their health of mind the hair recovers its smoothness.

Contraction of the Platysma Myoides Muscle.—This muscle is spread over the sides of the neck, extending downwards to a little beneath the collar-bones and upwards to the lower part of the cheeks. Its contraction draws the corners of the mouth and the lower parts of the cheeks downwards and backwards; producing at the same time divergent, longitudinal, prominent ridges on the sides of the neck in the young; and in old, thin persons fine transverse wrinkles.

It is strongly contracted under the influence of fear, and hence has been called the *muscle of fright.* Its contraction, however, is inexpressive unless associated with widely open eyes and mouth. Nevertheless, a man may exhibit extreme terror in the plainest manner by death-like pallor, by drops of perspiration on his skin, and by utter prostration, with all the muscles of his body, including the platysma, completely relaxed. The platysma muscle is also contracted in vomiting, nausea, disgust, and rage. A lady, an excellent musician, in singing high notes always contracts her platysma; so also does a certain young man, when playing the flute.

Whenever a person starts at any sudden sight or sound, he instantaneously draws a deep breath; and thus the contraction of the platysma may have become associated with the sense of fear. Again, the first sensation of fear, or the imagination of something dreadful, commonly excites a shudder. Now, as a shudder often accompanies the first sensation of fear, we have a clue to its action.

Dilatation of the Pupils.—The pupils are enormously dilated during terror. Dilatation of the pupil produced by fear has been observed in a water-spaniel, a retriever, a fox-terrier, and a cat. Pain also causes dilatation of the pupil. No doubt the fears of man have often been excited in the dark; but hardly so often or so exclusively as to account for a fixed and associated habit having thus arisen. It seems more probable that the brain is directly affected by the powerful emotion of fear and reacts on the pupils.

Horror.—The state of mind expressed by horror is in some cases almost synonymous with terror. Many a man must have felt, before the blessed discovery of chloroform, great horror at the thought of an impending surgical operation. We feel horror if we see a child exposed to some instant danger. Almost every one would experience the same feeling in witnessing a man being tortured. In these cases we put ourselves, in imagination, in the position of the sufferer and feel something akin to fear.

Sir C. Bell remarks that "horror is full of energy

(Fig. 22); the body is in the utmost tension, not unnerved by fear." It is therefore probable that horror would generally be accompanied by the strong contraction of the brows; but as fear is one of the

FIG. 22.—Horror and Pain. The expression has much in common with that of Terror, but the brow is *vertically* furrowed through the strong contraction of the two *Corrugators* and the general attitude is one of tension and extreme energy.

elements the eyes and mouth would be opened, and the eyebrows would be raised as far as the antagonistic action of the corrugators permitted. A tortured man, as long as his sufferings allowed him to feel any dread for the future, would probably exhibit horror in an extreme degree.

Horror is generally accompanied by various gestures, which differ in different individuals. The whole body is often turned away or shrinks; or the arms are violently protruded as if to push away some dreadful object. The most frequent gesture, as far as can be inferred from the acting of persons who endeavour to express a vividly imagined scene of horror, is the raising of both shoulders with the bent arms pressed closely against the sides or chest. These movements are nearly the same with those commonly made when we feel very cold; and they are generally accompanied by a shudder as well as by a deep expiration or inspiration, according as the chest happens at the time to be expanded or contracted. The sounds thus made are expressed by words like *uh* or *ugh.* It is not, however, obvious why, when we feel cold or express a sense of horror, we press our bent arms against our bodies. raise our shoulders, and shudder. I have noted that monkeys when cold huddle together, contract their necks, and raise their shoulders. Professor Gomperz of Vienna, however, has suggested that the gesture of pressing the folded arms to the sides may have been originally associated serviceably with the sensation of cold. This gesture would, therefore, become associated with the shuddering caused by cold. Thus, when a shudder is caused by the feeling of horror, the above gesture might accompany it simply because it had become "adherent" to it in the frequently recurring sensation of cold. It is not difficult to guess why the gesture with the arms should be associated with cold, since by flexing the arms and pressing them to the sides, the exposed surface is diminished.

Conclusion.—I have now endeavoured to describe the diversified expressions of fear, in its gradations from mere attention, to a start of surprise, to extreme terror and horror. Men, during numberless generations, have endeavoured to escape from their enemies by headlong flight, or by violently struggling with them; and such great exertions will have caused the heart to beat rapidly, the breathing to be hurried, the chest to heave, and the nostrils to be dilated. As these exertions have often been prolonged to the last extremity the final result will have been utter prostration, pallor, perspiration, trembling, or complete relaxation of all the muscles. And now, whenever the emotion of fear is strongly felt, though it may not lead to any exertion, the same results tend to reappear through the force of inheritance and association.

In the case of lower animals involuntary bristling of the hair serves, together with certain voluntary movements, to make them appear terrible to their enemies; and as the same involuntary and voluntary actions are performed by animals nearly related to man, we are led to believe that man has retained through inheritance a relic of them, though they are now useless. It is a remarkable fact that the minute unstriped muscles by which the hairs thinly scattered over man's almost naked body are erected should have been preserved to the present day; and that they should still contract under the same emotions—namely, terror and rage—which cause the hairs to stand on end in the lower members of the order to which man belongs.

CHAPTER XIII

SELF-ATTENTION—SHAME—SHYNESS—MODESTY—BLUSHING

BLUSHING is the most human of all expressions. Monkeys redden from passion, but no animal really blushes. The reddening of the face from a blush is due to the relaxation of the muscular coats of the small arteries by which the capillaries become filled with blood. No doubt if there be at the same time much mental agitation the general circulation will be affected; but it is not due to the action of the heart that the network of minute vessels covering the face becomes under a sense of shame gorged with blood. We can cause laughing by tickling the skin, weeping or frowning by a blow, trembling by the fear of pain, and so forth; but we cannot cause a blush by any action of the body. It is the mind which must be affected. Blushing is not only involuntary; but the wish to restrain it, by leading to self-attention, actually increases the tendency.

The young blush much more freely than the old, but not during infancy, though we know that infants at a very early age redden from passion. I have received authentic accounts of two little girls blushing at the ages of between two and three years; and of another sensitive child, a year older, blushing when reproved for a fault. It appears that the mental powers of infants are not as yet sufficiently developed to allow of their blushing. Hence, also, it is that idiots rarely blush, their faces flush from joy and from anger.

Women blush much more than men. It is rare to see an old man, but not nearly so rare to see an old woman, blushing. Laura Bridgman, born blind and deaf, blushes. Three children born blind, out of seven or eight in the Worcester College Asylum, are great blushers. The blind are not at first conscious that they are observed, and this knowledge has to be impressed on their minds as part of their education, and the impression thus gained would greatly strengthen the tendency to blush by increasing the habit of self-attention.

The tendency to blush is inherited. Dr. Burgess gives the case of a family consisting of a father, mother, and ten children, all of whom, without exception, were prone to blush to a most painful degree.

In most cases the face, ears, and neck are the sole parts which redden; but many persons whilst blushing intensely feel their whole bodies growing hot and tingling, which shows that the entire surface must be affected. Every one must have noticed how easily after one blush fresh blushes chase each other over the face. Blushing is preceded by a peculiar sensation in the skin. The reddening of the skin is generally succeeded by a slight pallor, which shows that the capillary vessels contract after dilating. In rare cases paleness is caused under conditions which would naturally induce a blush. For instance, a young lady told me that in a large and crowded party she caught her hair so firmly on the button of a passing servant that it took some time before she could be extricated; from her sensations she imagined that she had blushed crimson; but was assured by a friend that she had turned extremely pale.

Blushing in the Various Races of Man.—The small vessels of the face become filled with blood from the emotion of shame in almost all the races of man, though in the very dark races no distinct change of colour can be perceived. Blushing is evident in all the Aryan nations of Europe, and to a certain extent in those of India.

The Semitic races blush freely. It is said of the Jews in the Book of Jeremiah, "Nay, they were not at all ashamed, neither could they blush." According to Professor Robertson Smith these words do not imply blushing and it is possible that pallor is meant. There is, however, a word *haphar* occurring in Psalm xxxiv. 5, which probably means to blush. An Arab who was managing his boat clumsily, when laughed at by his companions, blushed to the back of his neck.

The Chinese blush, and have the expression "to redden with shame." The native Malays of the interior blush. In two Malays the face, neck, breast, and arms were observed to blush; and in a third Malay the blush extended to the waist.

Polynesians and New Zealanders blush freely. Forster says that "you may easily distinguish a spreading blush" on the cheeks of the fairest women in Tahiti. The Rajah Brooke has never observed the least sign of a blush with the Dyaks of Borneo; on the contrary, under circumstances which would excite a blush in us, they assert "that they feel the blood drawn from their faces." Humboldt quotes the sneer of the Spaniard concerning the Indians of South America: "How can those be trusted who know not how to blush?"

Moreau gives a detailed account of the blushing of a Madagascar negress-slave when forced by her brutal master to exhibit her naked bosom. The skin, perhaps, from being rendered more tense by the filling of the capillaries, would reflect a somewhat different tint from what it did before. That the capillaries of the face of the negro become filled with blood under the emotion of shame, we may feel confident; because an albino negress, described by Buffon, showed a faint tinge of crimson on her cheeks when she exhibited herself naked. Cicatrices of the skin remain for a long time white in the negro, and a scar of this kind on the face of a negress became red when she was charged with a trivial offence. Mulattoes are often great blushers, blush succeeding blush over their faces. From these facts there can be no doubt that negroes do blush, although invisibly. A blush has never been observed among Kaffirs of South Africa or Australians, but both races look down to the ground when ashamed.

These facts suffice to show that blushing, whether or not there is any change of colour, is common probably to all of the races of man.

Movements and Gestures which accompany Blushing.—Mr. Wedgwood says that the word shame "may well originate in the idea of shade or concealment, and may be illustrated by the Low German *scheme*, shade or shadow." Under a keen sense of shame there is a strong desire for concealment. We turn away the whole body, more especially the face, which we endeavour in some manner to hide. An ashamed person can hardly endure to meet the gaze of those present, so that he almost invariably casts down his eyes or looks askant. As there generally exists at the same time a strong wish to avoid the appearance of shame, a vain attempt is made to look direct at the person who causes this feeling; and the antagonism between these opposite tendencies leads to various restless movements in the eyes.

The aborigines in various parts of the world exhibit shame by looking downwards. Ezra cries out, "I am ashamed and blush to lift up my face." In Isaiah we meet with the words, "I hid not my face from shame." Seneca remarks "that the Roman players hang down their heads, fix their eyes on the ground and keep them lowered in acting shame." Shakespeare makes Marcus say to his niece, "Ah! now thou turn'st away thy face for shame." Little children, when shy or ashamed, turn away, and still standing up, bury their faces in their mother's gown, or they throw themselves face downwards on her lap.

Confusion of Mind.—Most persons whilst blushing intensely have their mental powers confused. This is recognized in such common expressions as "she was covered with confusion." Persons in this condition lose their presence of mind, utter singularly inappropriate remarks, are much distressed, stammer, and make awkward movements or strange grimaces.

With persons just *commencing* to blush it appears, judging from their bright eyes and lively behaviour, that their mental powers are somewhat stimulated. It is only when the blushing is excessive that the mind grows confused. It would seem that the capillaries of the face are affected during blushing before that part of the brain is affected on which the mental powers depend.

Conversely, when the brain is primarily affected the circulation of the skin is so in a secondary manner. In such cerebral cases, when the skin on the thorax or abdomen is gently touched, the surface becomes suffused in less than half a minute with bright red marks, called *cerebral maculæ*, which spread to some distance on each side of the touched point, and persist for several minutes. If, then, there exists, as cannot be doubted, an intimate sympathy between the capillary circulation in that part of the brain on which our mental powers depend and in the skin of the face, it is not surprising that the moral causes which induce intense blushing should likewise induce, independently of their own disturbing influence, much confusion of mind.

Mental States which induce Blushing.—These consist of shyness, shame, and modesty; the essential element in all being self-attention. Many reasons can be assigned for believing that originally self-attention directed to personal appearance, in relation to the opinion of others, was the exciting cause; the same effect being subsequently produced, through the force of association, by self-attention in relation to moral conduct. It is not the simple act of reflecting on our own appearance, but the thinking what others think of us, which excites a blush. In absolute solitude the most sensitive person would be quite indifferent about his appearance. We feel blame or disapprobation more acutely than approbation; and consequently depreciatory remarks and ridicule cause us to blush more readily than does praise. But undoubtedly praise and admiration are highly efficient: a pretty girl blushes when a man gazes intently at her, though she knows perfectly well that he is not depreciating her. Many children, as well as old and sensitive persons, blush when praised.

Attention directed to personal appearance, and not to moral conduct, has, I believe, been the fundamental element in the acquirement of the habit of blushing. Nothing makes a shy person blush so much as any remark on his personal appearance. One cannot notice even the dress of a woman much given to blushing without causing her face to crimson.

Women are much more sensitive about personal appearance than men are, especially elderly women in comparison with elderly men, and they blush much more freely. The young of both sexes are much more sensitive on this same head than the old, and they also blush much more freely than the old. Children at a very early age do not blush, nor do they show those other signs of self-consciousness which generally accompany blushing; and it is one of their chief charms that they think nothing about what others think of them. At this early age they will stare at a stranger with a fixed gaze and unblinking eyes, as on an inanimate object, in a manner which we elders cannot imitate.

It is plain to every one that young men and women blush incomparably more in the presence of the opposite sex than in that of their own. A young man, not very liable to blush, will blush intensely at any slight ridicule of his appearance from a girl whose judgment on any important subject he would disregard. No happy pair of young lovers, valuing each other's admiration and love more than anything else in the world, probably ever courted each other without many a blush. Even the barbarians of Tierra del Fuego blush in regard to women and also at their own personal appearance.

Of all parts of the body, the face is most considered and regarded, as is natural from its being the chief seat of expression and the source of the voice. It is also the chief seat of beauty and of ugliness, and throughout the world is the most ornamented. The face, therefore, will have been subjected during many generations to much closer and more earnest self-attention than any other part of the body; and we can therefore understand why it should be the most liable to blush. With Europeans the whole body tingles slightly when the face blushes intensely; and with the races of men who habitually go nearly naked the blushes extend over a much larger surface than with us. These facts are to a certain extent intelligible, as the self-attention of primeval man, as well as of the existing races which still go naked, will not have been so exclusively confined to their faces as is the case with the people who now go clothed.

It is probable that primeval man before he had acquired much moral sensitiveness would have been highly sensitive about his personal appearance, at least in reference to the other sex, and he would consequently have felt distress at any depreciatory remarks about his appearance; and this is one form of shame. And as the face is the part of the body which is most regarded, it is intelligible that any one ashamed of his personal appearance would desire to conceal this part of his body. The habit, having been thus acquired, would naturally be carried on when shame from strictly moral causes was felt; and it is not easy otherwise to see why under these circumstances there should be a desire to hide the face more than any other part of the body.

The habit of turning away, or lowering his eyes, or restlessly moving them from side to side, so general with every one who feels ashamed, probably follows from the conviction that he is being intently regarded; and he endeavours, by not looking at those present, and especially not at their eyes, momentarily to escape from this painful conviction.

Shyness.—This odd state of mind appears to be one of the most efficient of all the causes of blushing. Shyness is, indeed, chiefly recognised by the face reddening, by the eyes being averted or cast down, and by awkward, nervous movements of the body. Many a woman blushes from this cause, perhaps a thousand times to once that she blushes from having done anything of which she is truly ashamed. Shyness seems to depend on sensitiveness to the opinion of others with respect to external appearance. Strangers, who know nothing about our character, often criticize our appearance: hence shy persons are particularly apt to blush in the presence of strangers. The consciousness of anything peculiar in the dress or in the face makes the shy intolerably shy. On the other hand, in those cases in which conduct and not personal appearance is concerned, we are much more apt to be shy in the presence of acquaintances whose judgment we in some degree value, than in that of strangers. Some persons are so sensitive that the mere act of speaking to almost any one is sufficient to rouse their self-consciousness, and a slight blush is the result.

Disapprobation or ridicule, from our sensitiveness on this head, causes shyness and blushing much more readily than does approbation. The conceited are rarely shy; for they value themselves much too highly to expect depreciation. Persons who are exceedingly shy are rarely shy in the presence of those with whom they are quite familiar, and of whose good opinion and sympathy they are perfectly assured.

Shyness is closely related to fear, yet it is distinct from fear in the ordinary sense. A shy man no doubt dreads the notice of strangers, but can hardly be said to be afraid of them; he may be as bold as a hero in battle, and yet have no self-confidence about trifles in the presence of strangers. Almost every one is extremely nervous when first addressing a public assembly, and most men remain so throughout their lives; but this appears to depend on the consciousness of an impending and perhaps unaccustomed exertion rather than on shyness, although a timid or shy man no doubt suffers on such occasions infinitely more than another. The "abashed" feeling experienced at such times as well as the *stage-fright* of actors is attributable to simple apprehension or dread.

Shyness comes on at a very early age. As it apparently depends on self-attention, we can perceive how right are those who maintain that reprehending children for shyness, instead of doing them any good, does much harm, as it calls their attention still more closely to themselves. Nothing hurts young people more than to have their countenances scrutinized. Under the constraint of such examinations they can think of nothing but that they are looked at, and feel nothing but shame or apprehension.

Moral Causes.—Blushing from strictly moral causes is likewise related to regard for the opinion of others. It is not the consciousness of guilt, but the thought that others think or know us to be guilty which crimsons the face. A man may feel thoroughly ashamed of having told a falsehood, without blushing; but if he is detected he will instantly blush, especially if detected by one whom he reveres.

On the other hand, a man may be convinced that God witnesses all his actions, and he may feel deeply conscious of some fault and pray for forgiveness; but this will not excite a blush. The explanation of this difference between the knowledge by God and man of our actions lies, I presume, in the association between man's disapprobation of our immoral conduct and his depreciation of our personal appearance, so that both lead to similar results, whereas the disapprobation of God brings up no such association.

Many a person has blushed intensely when accused of some crime, though completely innocent of it. Even the thought that others think that we have made a stupid remark is amply sufficient to cause a blush, although we know that we have been misunderstood. An action may be meritorious or of an indifferent nature, but a sensitive person, if he suspects that others take a different view of it, will blush. For instance, a lady by herself may give money to a beggar without blushing, but if others are present, and she suspects that they think her influenced by display, she will blush.

Breaches of Etiquette.—Breach of the laws of etiquette—that is, any impoliteness, *gaucherie*, impropriety, or inappropriate remark—though quite accidental, will cause the most intense blushing of which a man is capable. Even the recollection of such an act after an interval of many years will make the whole body to tingle. So strong, also, is the power of sympathy that a sensitive person will sometimes blush at a flagrant breach of etiquette by a perfect stranger, though the act may in no way concern her.

Modesty.—This is another powerful agent in exciting blushes. Blushing here has the usual signification of regard for the opinion of others. But modesty frequently relates to acts of indelicacy; and indelicacy is an affair of etiquette, as we clearly see with the nations that go altogether or nearly naked. He who is modest, and blushes easily at acts of this nature, does so because they are breaches of a firmly and wisely established etiquette. This is, indeed, shown by the derivation of the word *modest* from *modus*—a measure or standard of behaviour. A blush due to this form of modesty is, moreover, apt to be intense, because it generally relates to the opposite sex; and we have seen how in all cases our liability to blush is thus increased.

The fact that blushes may be excited in absolute solitude seems opposed to the view that the habit originally arose from thinking about what others think of us. Shakespeare may have erred (or he may have meant that the blush was unseen, not that it was absent), when he made Juliet say to Romeo:

> "Thou know'st the mask of night is on my face;
> Else would a maiden blush bepaint my cheek,
> For that which thou hast heard me speak to-night."

Finally, blushing, whether due to shyness, to shame or to modesty, depends in all cases on a sensitivity regarding the opinion, especially if depreciative, of others, primarily in relation to our personal appearance, and secondarily, through the force of association and habit, in relation to the opinion of others on our conduct.

Theory of Blushing.—Why should the thought that others are thinking about us affect our capillary circulation? Sir C. Bell insists that blushing "is a provision for expression, as may be inferred from the colour extending only to the parts most exposed. Dr. Burgess believes that it was designed by the Creator in "order that the soul might have sovereign power of displaying in the cheeks the various internal emotions of the moral feelings," so as to serve as a check on ourselves, and as a sign to others that we are violating rules which ought to be held sacred. Gratiolet merely remarks that the phenomena of pallor and of blushing that distinguish man are a natural sign of his high perfection.

The belief that blushing was *specially* designed by the Creator is opposed to the general theory of evolution which is now so largely accepted. Those who believe in design will find it difficult to account for shyness being so frequent and efficient a cause of blushing, as it makes the blusher to suffer and the beholder uncomfortable, without being of the least service to either. They will also find it difficult to account for negroes and other dark-coloured races blushing, in whom a change of colour in the skin is scarcely or not at all visible.

No doubt a slight blush adds to the beauty of a maiden's face; and the Circassian women who are capable of blushing fetch a higher price in the seraglio of the Sultan than less susceptible women. But the firmest believer in the efficacy of sexual selection will hardly suppose that blushing was acquired as a sexual ornament.

The hypothesis which appears to me the most probable is that attention closely directed to any part of the body tends to interfere with the ordinary tonic contraction of the small arteries of that part. These vessels, in consequence, become at such times more or less relaxed, and are instantly filled with arterial blood. This tendency will have been much strengthened if frequent attention has been paid during many generations to the same part, owing to nerve-force readily flowing along accustomed channels through the power of inheritance. Assuming that the capillary vessels can be acted on by close attention, those of the face will have become eminently susceptible. Through the force of association, the same effects will tend to follow whenever we think that others are considering or censuring our actions or character.

Certain glands are much influenced by thinking of the conditions under which they have been habitually excited. This is familiar to every one in the increased flow of saliva, when the thought, for instance, of intensely acid fruit is kept before the mind. An earnest and long-continued desire either to repress, or to increase, the action of the lacrymal glands is effectual. Some curious cases have been recorded of the power of the mind on the mammary and uterine functions. Dr. J. Crichton Browne is convinced that attention directed for a prolonged period on any part or organ may ultimately influence its capillary circulation and nutrition. He refers to a married woman fifty years of age who laboured under the firm and long-continued delusion that she was pregnant. When the expected period arrived she acted precisely as if she were being really delivered of a child, and seemed to suffer extreme pain, so that the perspiration broke out on her forehead. The result was that a state of things returned, continuing for three days, which had ceased during the six previous years.

Dr. Maudsley quotes some curious statements with respect to the improvement of the sense of touch by practice and attention. When we direct our whole attention to any one sense its acuteness is increased, and the continued habit of close attention, as with blind people to that of hearing and with the blind and deaf to that of touch, improves the sense in question permanently.

Lastly, some physiologists maintain that the mind can influence the nutrition of parts. Sir J. Paget has given a curious instance of the power, not, indeed, of the mind, but of the nervous system, on the hair. A lady "who is subject to attacks of what is called nervous headache always finds in the morning after such an one that some patches of her hair are white, as if powdered with starch. The change is effected in a night, and in a few days after the hairs gradually regain their dark brownish colour." A London physician who suffers from neuralgia over the eye-brow sustains with each attack whitening of a patch of hair in the brow, which, however, recovers its colour again when the attack is past.

We thus see that close attention certainly affects various parts and organs which are not properly under the control of the will. When, therefore, we voluntarily concentrate our attention on any part of the body, the cells of the brain which receive impressions from that part are, it is probable, stimulated into activity. This may account, without any local change in the part to which our attention is earnestly directed, for pain or odd sensations being there felt or increased.

The power of attention to a part of the body seems to depend on the vaso-motor system being affected in such a manner that more blood is allowed to flow into the capillaries of the part in question. This increased action of the capillaries may be combined with the simultaneously increased activity of the sensorium.

Now, as men during endless generations have had their attention directed to their personal appearance and especially to their faces, any incipient tendency in the facial capillaries to be thus affected will have become in the course of time greatly strengthened through nerve-force passing readily along accustomed channels. This appears to me to throw light on the leading phenomena connected with the act of blushing.

Chapter XIV

CONCLUDING REMARKS AND SUMMARY

I have now attempted to explain the origin of expressive actions in man and lower animals through three great principles: (1) movements which are serviceable if often repeated become habitual; (2) a habit of voluntarily performing opposite movements under opposite impulses has become firmly established in us; (3) the direct action of the excited nervous system on the body.

The frantic and senseless actions of an enraged man may be attributed in part to the undirected flow of nerve-force, and in part to the effects of habit. They pass into gestures included under our first principle; as when an indignant man unconsciously throws himself into a fitting attitude for attacking his opponent, though without any intention of making an actual attack. We see also the influence of habit in all the emotions and sensations which are called exciting; for they have assumed this character from having habitually led to energetic action. Whenever these emotions or sensations are slightly felt by us our whole system is disturbed through habit and association. Other emotions and sensations, such as those of pain, grief, and fear, are called depressing, because they have not habitually led to energetic action, but rather to exhaustion; they are consequently expressed chiefly by negative signs and by prostration. Other emotions, such as that of affection, do not commonly lead to action of any kind, and consequently are not exhibited by any strongly marked outward signs.

On the other hand, many of the effects due to the excitement of the nervous system seem to be quite independent of the flow of nerve-force along the channels which have been rendered habitual by former exertions of the will. Such effects, which often reveal the state of mind of the person thus affected, cannot at present be explained; [1] for instance, the change of colour in the hair from extreme terror or grief, the cold sweat and the trembling of the muscles from fear, the modified secretions of the intestinal canal, and the failure of certain glands to act.

Actions of all kinds, if regularly accompanying any state of mind, are at once recognized as expressive. These may consist of movements of any part of the body, as the wagging of a dog's tail, the shrugging of a man's shoulders, the erection of the hair, the exudation of perspiration, the state of the capillary circulation, laboured breathing, and the use of vocal or other sound-producing instruments. Even insects, by their stridulation, express anger, terror, and love.

We are so familiar with the fact of young and old animals displaying their feelings in the same manner that we hardly perceive how remarkable it is that a young puppy should wag its tail when pleased, depress its ears, and uncover its canine teeth when pretending to be savage, just like an old dog; or that a kitten should arch its little back and erect its hair when frightened and angry, like an old cat. When, however, we turn to less common gestures in ourselves, such as shrugging the shoulders as a sign of impotence, or the raising the arms with open hands and extended fingers as a sign of wonder, we feel perhaps too much surprise at finding that they are innate.

Certain other gestures apparently have been learnt like the words of a language. This seems to be the case with the joining of the uplifted hands and the turning up of the eyes in prayer.

But the far greater number of the movements of expression, and all the more important ones, are

[1] Such effects are now known to be brought about in great measure by the action of complex chemical products (hormones) liberated by the endocrine glands (adrenals, pancreas, pituitary, etc.).—C. M. B.

innate or inherited, and cannot be said to depend on the will of the individual.

The power of communication between the members of the same tribe by means of language has been of paramount importance in the development of man; and the force of language is much aided by the expressive movements of the face and body. Even infants, if carefully attended to, find out at a very early age that their screaming brings relief, and they soon voluntarily practise it. We may frequently see a person voluntarily raising his eyebrows to express surprise, or smiling to express pretended satisfaction and acquiescence. A man often wishes to make certain gestures conspicuous or demonstrative, and will raise his extended arms with widely opened fingers above his head to show astonishment, or lift his shoulders to his ears to show that he cannot or will not do something. The tendency to such movements will be strengthened or increased by their being thus voluntarily and repeatedly performed; and the effects may be inherited.

Movements at first used by only a few individuals to express a certain state of mind may sometimes have spread to others, and ultimately have become universal, through the power of imitation. There exists in man a strong tendency to imitation, independently of the conscious will. This is exhibited in certain brain diseases, and has been called the "echo sign." Patients thus affected imitate, without understanding, every gesture which is made and every word which is uttered even in a foreign language near them. In the case of animals the jackal and wolf have learnt under confinement to imitate the barking of the dog.

Monkeys soon learn to distinguish not only the tones of voice of their masters, but also the expression of their faces. Dogs well know the difference between caressing and threatening gestures or tones; and they seem to recognize a compassionate tone. But as far as I can make out, after repeated trials, they do not understand any movement confined to the features, excepting a smile or laugh; and this they appear, at least in some cases, to recognize. This limited amount of knowledge has probably been gained both by monkeys and dogs through their learning to associate harsh or kind treatment with our actions. Children, no doubt, would soon learn the movements of expression in their elders in the same manner as animals learn those of man. Moreover, when a child cries or laughs he knows in a general manner what he is doing and what he feels; so that a very small exertion of reason would tell him what crying or laughing meant in others.

The fact that all the chief expressions exhibited by man are the same throughout the world, affords a new argument in favour of the several races being descended from a single parent-stock, which must have been almost completely human in structure and in mind before the period at which the races diverged from each other.

We may confidently believe that laughter, as a sign of pleasure or enjoyment, was practised by our progenitors long before they deserved to be called human; for very many kinds of monkeys, when pleased, utter a reiterated sound, clearly analogous to our laughter. We may likewise infer that fear was expressed from an extremely remote period in almost the same manner as it now is by man.

Suffering, if great, will from the first have caused screams or groans to be uttered, the body to be contorted, and the teeth to be ground together. But our progenitors will not have exhibited those highly expressive movements of the features which accompany screaming and crying, until their circulatory and respiratory organs, and the muscles surrounding the eyes, had acquired their present structure. Weeping probably came on rather late in the line of our descent; and this conclusion agrees with the fact that our nearest allies, the anthropomorphous apes, do not weep. But as certain monkeys weep, this habit might have developed long ago in a sub-branch of the group from which man is derived. Our early progenitors, when suffering from grief or anxiety, would not have drawn down the corners of their mouths until they had acquired the habit of endeavouring to restrain their screams. The expression, therefore, of grief and anxiety is eminently human.

Rage will have been expressed at a very early period by threatening gestures, reddening of the skin, and by glaring eyes. From the fact that a frown serves as a shade in difficult and intent vision, it seems probable that this expression would not have become habitual until man had assumed a completely upright position, for monkeys do not frown when exposed to a glaring light. Our early progenitors, when enraged, would probably have exposed their teeth more freely than does present-day man. When angry they would not have held their heads erect, opened their chests, squared their shoulders, and clenched their fists, nor would they have practised the antithetical gesture of shrugging the shoulders as a sign of impotence or of patience, until they had acquired the upright attitude of man, and had learnt to fight with their fists. For the same reason, astonishment would not then have been expressed by a widely open mouth and by raising the arms and hands. Disgust would have been shown at a very early period by movements round the mouth, like those of vomiting, but the more refined manner of showing contempt or disdain would not have been acquired until a much later period.

Of all expressions, blushing seems to be the most strictly human, and it probably originated at a very late period in the long line of our descent.

The movements of expression in the face and body are of much importance for our welfare. They serve as the first means of communication between the mother and her infant; she smiles approval, or frowns disapproval, and thus encourages her child on the right path. We readily perceive sympathy in others by their expression; our sufferings are thus mitigated, our pleasures increased, and mutual good feelings strengthened. The movements of expression give vividness and energy to our spoken words; they may, and often do, reveal the thoughts of others more truly than do words, which may be falsified. The free expression by outward signs of an emotion intensifies it. Passions can be produced by putting hypnotized people in appropriate attitudes. On the other hand, the repression of all outward signs softens our emotions. Shakespeare well knew that the simulation of an emotion tends to arouse it in our minds:

> "Is it not monstrous that this player here,
> But in a fiction, in a dream of passion,
> Could force his soul so to his own conceit,
> That, from her working, all his visage wann'd;
> Tears in his eyes, distraction in 's aspect,
> A broken voice, and his whole function suiting
> With forms to his conceit? And all for nothing!"

The study of the expression of man's emotions confirms to a certain limited extent the conclusion that he is derived from some lower animal form, and supports the belief of the unity of the several races; but, as far as my judgment serves, such confirmation was hardly needed. Expression in itself, or the language of the emotions, is certainly of importance for the welfare of mankind.

our pleasures increased, and mutual good feelings strengthened. The movements of expression give vividness and energy to our spoken words; they may, and often do, reveal the thoughts of others more truly than do words, which may be falsified. The free expression by outward signs of an emotion intensifies it. Passions can be produced by putting hypnotized people in appropriate attitudes. On the other hand, the repression of outward signs softens our emotions. Shakespeare well knew that the simulation of an emotion tends to arouse it in our minds:

> "Is it not monstrous that this player here,
> But in a fiction, in a dream of passion,
> Could force his soul so to his own conceit,
> That, from her working, all his visage wann'd,
> Tears in his eyes, distraction in's aspect,
> A broken voice, and his whole function suiting
> With forms to his conceit? And all for nothing!"

The study of the expression of man's emotions confirms to a certain limited extent the conclusion that he is derived from some lower animal form, and supports the belief of the unity of the several races; but as far as my judgment serves, such confirmation was hardly needed. Expression in itself, or the language of the emotions, is certainly of importance for the welfare of mankind.

INDEX

(Figures in italics refer to illustrations.)

AARD WOLF, hair erection in, 41
Abbott, C. C., 13
Abstraction, 114
Affection, attitude, *15*, 16, *18*
Affirmation, gestures, 7, 136–139
Agony, 26, 28. *See also* Pain.
Alarm, 29
Alertness, *30*
Amazement, 140–142
Anger, 16, *17*, *19*, *20*, 34–36, *40*, *41*, *49*, 66, 67, 114, 115, 122, 136. *See also* Rage.
Antithesis, principle of, 4, 20–23, 135, 136, 143, 170
Anubis baboon, 65
Anxiety, 170
Arrectores pili, 42–44
Astonishment, 36–38, 70, *71*, 140–145
Atavistic traits, 121
Attention, 53, 56, *70*, *71*, 140, 142

BABOON, exposure of teeth, *52*, 67, 125
——, hair erection, 40
Back, arching of, *20*, 60, 61
——, thrill down the, 25, 108
Bain, Alexander, 6, 105, 147
Barbary ape, 65
Bark, the, 29, 35, 168
Bartlett, Mr., 13, 51, 79
Baudry, M., 55, 57, 136
Beagle, H.M.S., 143
Bear, shading of eyes, 112, 113
Bee, sound of, 39
Beer, Prof., 10
Bell, Sir Charles, 1–3, 56, 83, 110, 124, 149, 163
Belly, stroking the, 104
Blair, Rev. H. H., 113
Blind, blushing among, 154
Blushing, 26, 119, 153–156, 159–161
——, a human trait, 170
——, causes of, 157, 162, 163, 165
——, first appearance, 153, 158
Boar, retraction of ears, 51
Body, effect of mind on, 6, 164, 165
Bond, F. W., 30, 52
Breath, holding the, 26, 69, 117
Brehm, A. E., 40
Bridgman, Laura, 98, 135, 137, 153
Brooke, the Rajah, 155
Brown, Dr. R., 47
Browne, Dr. Crichton, 88, 90, 101, 164
Buffon, G. Le Clerc, 155
Bull, anger, 62
——, exposure of teeth, 55
Bulmer, Mr. J., 125
Burgess, Dr., 119, 154, 163

Callithrix sciureus, 66
Camel, fighting attitude, 50
Canis latrans, 35
Caracal, *19*
Cat, affection, *18*, 60
——, alertness, *30*
——, anger, *17*, 59
——, kneading habit, 60

Cat, threatening attitude, 16, *17*, *20*
——, terror, 60, 61
Caterpillar, fatal habit of, 5
Cattle, emotion, 50
——, pain, 33
Cebus monkey, emotions, 65, 66
——, sneezing of, 77
Cerebral maculae, 157
Cercopithecus, emotion, 51, 67
Chacma baboon, yawning, *52*
Chameleon, inflation, 44
Chamois, stamping, 62
Chaucer, 75
Cheetah, purring, 61
Chimpanzee, astonishment, 142
——, hair erection, 39
——, kissing, 106
——, laughter, 63
——, pouting, *68*, 69, 115
——, smile, *64*
——, sneezing, 77
Clapping the hands, 29, 98, 103
Cleland, Prof., 127
Clenching the fists, 119, 122, 123, 136, 147, 170
"Cold shoulder," the, 116
"Cold sweat," the, 146, 167
Colour change in emotion, 26, 27, 31, 67, 87, 118, 119, 122, 146, 152–155, 163, 165, 167, 170
Comrie, Dr., 121
Contempt, 118, 127, *128*, 129, 132, 133. *See also* Disdain.
Cooke, T. P., 124
Corrugator muscles, *72*, 88, 89, 92, 93, 110, 113, *150*
Cow, emotion in, 54
Creation. *See* Special C.
Crying, 169
Cynocephalus baboon, yawning, *52*
Cynopithecus, emotion, 65
Cynopithecus niger, 53, 67, 70

Daily Mail, 64
Darwin, Charles, frontispiece, v. 24, 33, 50, 60
Day, Mr. F., 39
Deaf and dumb gestures, 21
Deceit, 133
Deer, fighting, 51
Depressor muscles, 95, 96
Derision, 105
Determination, 35, 114–117, 119, 135
Devotion, attitude, 108, 109, 167
Dickens, Charles, 121
Disdain, 118, *128*, 133. *See also* Contempt.
Disgust, 127–133, *130*, 149
——, first appearance, 170
Dog, affection, *15*, 54, 55
——, bark, 35
——, grin, 55, 56
——, hostile attitude, *14*
——, impatience, 36
——, pain, 56
——, snarl, *49*, 55
——, terror, 56, 57
——, "turning round" habit, 10, 11
Donders, Prof., 76, 77
"Down in the mouth," *94*, 95, 170
Duchenne, Dr., 63, *71*, 90, 91, 96
Dumont, L., 100
Dunbar, J. B., 33

Ears, retraction, 4, 12, 16, *17*, *20*, 22, 31, 48–51, *49*, 53, 54, 58, 59, 61, 62, 67, 142
Echo sign, 168
Elephant, fighting, 51
——, tears, 79
Emotions and music, 25, 36, 37, 108
Endocrine glands, v. 167
Esquimaux dogs, 11
Excrement covering, 11, 12
Exophthalmos, 77

Eyeballs, protrusion of, 76, 77, 120, 147
Eyes, "screwing up" the, *82*, 92
——, shading the, 110–113

Face, why the site of blushing, 157–159, 165, 167
Fear, 1, 31, 38, 70. *See also* Terror, Fright.
Feathers, erection, 39–43, *40*, *41*, 61
Feeling and action, 2
Fennec, "turning round" habit, 11
Fighting attitudes. *See* Horse, Giraffe, etc.
Finger, shaking the, 139
Fish, erection of spines, 39
Fists, clenching of, 119, 122, 123, 136, 147, 170
Flamingo, patting the ground, 13
Fox, backward scratching, 12
——, expression, 59
Fra Angelico, 90
Fright, muscle of, *145*, 148. *See also* Terror.
——, stage, 161
Frog, inflation, 44
——, reflex action in headless, 8
Frontal muscles, 89, 92–94, 104
Frown, the, 69, 70, 72, 87, 111–114, 120–122, 124, 136
Frowning, origin, 110–112, 170

Gaskell, Mrs. E. C., 73
Gibbon ape, call of, 36
Gill, Mr. Wyatt, 106
Giraffe, fighting, 50
Glands and the emotions, 25, 28, 84, 147, 164, 167
Globus hystericus, 88
Goat, fighting, 50
Goldfinch, erection of feathers, 41, 42
Gomperz, Prof., 144, 151
"Goose-skin," 42, 43
Gorilla, erection of hair, 40
——, pouting, 115
——, rage, 70
Gratiolet, Pierre, 163
Grief, 87
——, a human trait, 170
——, muscles, 88–94
Grin, the, 55, 56, 58, 121
Guanacoe, fighting, 50
Guenon monkey, fighting, 67
Gunning, Dr., 77
Gunther, Dr. A. C. L., 44, 47

Habit, 4–7, 27, 30, 34, 43, 48, 55, 74, 75, 85, 94, 96, 97, 108, 111, 112, 159, 166
Hair, erection of, *14*, 16, *20*, 26, 35, 39–43, 48, 61, 62, 71, 100, 120, 148, 152, 167
Hare, cry of, 33
Hartshorne, B. F., 103
Hate, 118
Haughtiness, 127, 134
Head, nodding, 7, 136–140
——, shaking, 136–139
Heart during emotion, 25–29, 57, 119, 122, 141, 146, 147, 152, 153
Helmholtz, Prof. H. L., 37, 38
Helplessness, 21, 134–136
Hen, anger, *40*
Henle, 77
High spirits, 104
Hinton, C., 99
Hippopotamus, pain, 26
——, retraction of ears, 50
Holbeach, H., 127
Homer, 98, 120
Hormones, v, 167
Horror, 7, 149, *150*, 151
Horse, anger, 61
——, cry of, 33
——, fear, 61, 62
——, fighting, 50
——, nibbling, 12
——, play of, 58
——, poise of tail, 54
Humboldt, Friedrich, 66

Huxley, T. H., 6, 115
Hyæna, hair erection, 41
——, kneeling, 57, 58
Hypnotism and emotions, 171

Idiots, laughter of, 98
Impatience, 13, 36, 61
Inflation of body, 44, 45
Insects, sounds of, 39, 167
Intelligence, apparent, 8
Iris, movements of, 10, 81. *See also* Pupil.

Jackal, backward scratching, 12
——, bark, 35
——, expression, 58, 59
——, "turning round" habit, 11
Jealousy, 133
Joy, 29, 98, 104

Kagu, patting the ground, 13
Kangaroo, fighting, 51
King, Major Ross, 51
Kissing, 106

Lacrymal glands, 73, 80, 81, 83–85, 106, 107, 164
Laughter, 36, 37, 63, 98–101, 103–105
——, causes of, 98, 99
——, origin, 127, 169
——, paroxysm, 102, 103
——, tears, 78
Leichardt, 104
Lewes, G. H., 10
Licking habit, 55, 59, 91
Lion cub, hair erection, 41
——, playfulness, *31*
——, rage, 34, 35
Lioness, maternal love, *31*
Lips, muscles, 73
——, protrusion, *68*, 69
Lizard, tail movements, 59, 60
Lloyd, R. M., 59
"Looking down on" expression, 127, *128*, 134

Louis XVI., 118
Love, 105
Lubbock, Sir J., 106
"Lump in the throat," 88
Lynx, anger, *19*
——, arching of back, 61

Macacus monkey, emotion, 53, 67
Macalister, Prof., 69
Mammary glands, kneading, 13
Mandrill, colour changes, 67
Martin, W. L., 67
Maternal love, 30, *31*
Matthews, W., 143
Maudsley, Dr. H., 8, 121, 122, 164
Midas œdipus, erection of hair, 40
Mind and body, 6, 163–165
——, confusion, 156, 157
Modesty, 162
Monkeys, call of, 36, 37
——, colour changes, 67, 153
——, expression, *52*, 63, *64*, *68*
——, fighting, 51, 67
——, incipient laughter, 169
——, weeping, 169
Moose deer, fighting, 51
Moreau, M., 155
Moseley, H. N., 123, 137
Mosso, M., 25, 26
Mouth, depressors, *94*, 95, 96, 170
——, frothing, 120
——, hand over, 144
Müller, Prof. Max, 115
Munby, A. J., 146
Muscles of fright, *145*, 148
——, of grief, 88–94
——, of patience, 135
——, of pride, 134
——, of snarling, 124
See also Corrugator, Depressor, Frontal, Lips, Orbicular, Panniculus, Platysma, Pyramidal, Zygomatic.

Musculus superbus, 134
Music and emotion, 25, 36, 37, 108
Musk ox, stamping, 62

Nature, 13
Negation, 35, 136–139
Nervous system and expression, 5
Nicols, Arthur, 59
Nodding, 7, 136–140
Nose, "turning up the," 132
Nostrils, contraction, 128, 129
——, dilatation, 26, 27, 30, 56, 62, 70, 119, 120, 122, 152

Ocelot, purring, 61
Orang, astonishment, 142
——, call, 115
——, chuckling, 63
——, hair erection, 39
——, holding breath, 69
——, pouting, 68, 115
——, sneezing, 77
Orbicular muscles, 72, 75, 78, 79, 83, 84, 88, 89, 92, 93, 102, 103, 105, 114
Otaria pusilla, 50

Paget, Sir James, 165
Pain, 56, 66, 72, 149, 169
——, expression, 27, 31, *150*
Panniculus carnosus, 42, 44
Parsons, J., 123
Patience muscles, 135
Patting the ground, 13, 67. *See also* Stamping.
Pawing, 12, 61, 62
Perplexity, 6, 114
Petherick, John, 104
Piderit, Dr. 111, 121
Pitch, 136
Platysma muscle, *145*, 148, 149
Plautus, 114
Pointing, 5, 11
Porcupine, rattling, 38, 39, 48
Pouting, 66–69, *68*, 115, 116, 120, 132, 142, 143
——, origin, 116
Pregnancy, pseudo, 164
Pride, 127, 134
——, muscle, 134
Puff adder, inflation, 45
—— ——, striking, 9
Puma, purring, 61
——, tail lashing, 59
Pupils, contraction, 120
——, dilatation, 147, 149. *See also* Iris.
Purring, 16, 61
Pyramidal muscles, 88, 89, 92–94, 114

Rabbit, cry, 33
——, fighting, 51
——, stamping, 38
Rage, 118–120, 148, 149, 166, 170. *See also* Anger.
Rational *v.* teleological explanations, 3
Rattling, 46, 47
Reade, Winwood, 106, 140, 143, 144
Reeks, Henry, 112
Reflex action, 7–10, 42, 80, 81, 100, 131
Regurgitation, voluntary, 132
Rejlander, Mr., 127
Rengger, 66
Reynolds, Sir J., 103
Rhinoceros, fighting, 51
Rice ordeal, 147

St. John, Mr., 13
Scorn, 129, 133
Scott, J., 21, 91
——, Sir W., 56
Scratching, 6, 11, 12
Screaming, 71–78, 84, 85, 93–97, 99, 102, 112, 120, 147, 169, 170
——, uses, 37, 168
Sculptors, errors of, 90
Seal, retraction of ears, 50
Seneca, 156

Sensation, 4
Shakespeare, 132, 133, 140, 156, 162, 171
——, on rage, 120
Shaking the head, 136–139
Shame, 26, 105, 156, 159, 160
Shawe, Dr. T. C., 44
Sheep, anger, 62
——, call, 34
——, fighting, 50
Sheldrake, patting the ground, 13
Shrugging, 21, *57*, 134–136, 167
Shudder, 149–151
Shyness, 105, 115, 156–162
Slyness, *95*, 134
Smile, 63, *64*, 98, 99, 101, 102, 104
—— of contempt, 127
——, sardonic, 125
Smith, Prof. Robertson, 155
Snake, rattling, 39
——, tail lashing, 59, 60
Snapping the fingers, 129
Snarl, *49*, 121, 124, 125
Snarling muscles, 124
Sneering, 123–125
Sneezing, 7, 8, 75, 77, 81, 83, 86
Sobbing, 88
——, a human trait, 75
Somerville, William, 56
Sophocles, Mr., 138
Sounds, emission of, 33, 39, 46, 47, 167
Special creation and emotions, 2, 3, 163
Spencer, Herbert, on fear, 1
——, ——, on feeling, 2, 4
——, ——, on voice, 35
Spenser, Edmund, 133
Spitting, 16, 50, 131–133
Stack, Rev. J. W., 122
Stage fright, 161
Stamping, 38, 62, 98. *See also* Patting.
"Starting" habit, 9, 10, 37, 61, 142, 149
Stork, clattering, 39
Strabo, 129
Stridulation, 167
Stuart, Mr., 141
Sulkiness, *68*, 69, 115, 116, 135
Surprise, *70*, *71*, 140–144
Suspicion, *30*, 133
Sutton, Mr., *71*, *77*
Swallowing, 9
Swan, anger, *41*
Sympathy, 107, 108
Tail, lashing, 16, 22, 59
——, poise, *14*, *15*, 16, *17*, *18*, *20*, 21, 22, 54, 57–59
——, wagging, 4, 16, 21, 22, 59, 167
Tapaya douglasii, 44
Tears, 28, 66, 73–87, 97, 101–108
——, cause of, 77
——, first appearance, 73
——, function, 80, 81
——, in laughter, 101, 103
Teeth, exposure of, 15, 16, *17*, *19*, 31, 41, *49*, 50, *52*, 55, 56, 61, 67, *71*, 101, 115, 121, 123, *124*, 125, 126, 170
——, ——, in bull, 55
——, ——, in horse, 12
——, gnashing, 26, 51
——, grinding, 62, 86, 119, 169
Tenderness, 105, 106
Tennent, Sir E., 79
Tennyson, 119
Terror, 26, 32, 56, 57, 70, 71, 140, 144–149, *145*
——, in dog, 56, 57
——, in horse, 61, 62
Threatening gesture, ii, 16, *20*
Thumb biting, 129
Tickling, 42, 99, 100, 153
Trembling, 24, 25, 28, 38, *101*, 120, 147, 148, 152, 167
Turks, gestures of, vi, 137, 138

" Turning round " habit, 10, 11
" Turning up the nose," 132
Tylor, E. B., 36, 109

Vesalius, 138
Vinci, Leonardo Da, 143
Vocal cries, 33–39
—— organs, origin, 34, 35
Vomiting, 76, 131, 132, 149, 170

Wallace, A. R., 54, 63, 142, 144
Weddahs, absence of laughter, 103
Wedgwood, H., 109, 119, 156
Weeping, 72–86, 112
——, origin, 169
Whistling, 62, 143
Whitmee, Rev. S. J., 39
Winking, 70, 139
——, first appearance, 9
——, use of, 9, 10
Wolf, expression, 58, 59
——, backward scratching, 12

Yawning, *52*, 78, 83, 147

Zeki, Mr. Sedat, vi, 138
Zygomatic muscles, 101, 105